T0212116

Lecture Notes of the Institute for Computer Sciences, Social Informatics and Telecommunications Engineering 373

More information about this series at http://www.springer.com/series/8197

Maggie Cheng · Peng Yu ·
Yuan Hong · Huibin Jia (Eds.)

Smart Grid and Innovative Frontiers in Telecommunications

5th EAI International Conference, SmartGIFT 2020
Chicago, USA, December 12, 2020
Proceedings

Springer

Editors
Maggie Cheng 🄳
Illinois Institute of Technology
Chicago, IL, USA

Yuan Hong 🄳
Illinois Institute of Technology
Chicago, IL, USA

Peng Yu 🄳
Beijing University of Posts and
Telecommunications
Beijing, China

Huibin Jia 🄳
North China Electric Power University
Huabei, China

ISSN 1867-8211 ISSN 1867-822X (electronic)
Lecture Notes of the Institute for Computer Sciences, Social Informatics
and Telecommunications Engineering
ISBN 978-3-030-73561-6 ISBN 978-3-030-73562-3 (eBook)
https://doi.org/10.1007/978-3-030-73562-3

This Springer imprint is published by the registered company Springer Nature Switzerland AG
The registered company address is: Gewerbestrasse 11, 6330 Cham, Switzerland

Preface

We are delighted to introduce the proceedings of the fifth edition of the European Alliance for Innovation (EAI) International Conference on Smart Grid and Innovative Frontiers in Telecommunications (SmartGIFT 2020). This conference brought together researchers, developers and practitioners from around the world who are leveraging and developing smart grid technology for providing stable and reliable control and operation of power systems.

Due to safety concerns and travel restrictions caused by COVID-19, SmartGIFT 2020 took place online in a live stream on December 12, 2020. The technical program of SmartGIFT 2020 consisted of 11 full papers with 3 technical sessions. The technical sessions. Session 1 - Communications, Networks and Services; Session 2 - Security and Stable Control; Session 3 - Internet of Power Things and Big Data. Aside from the high-quality technical paper presentations, the technical program also featured a keynote speech given by Prof. Jianhui Wang from Southern Methodist University.

Coordination with the Steering Committee, Imrich Chlamtac, Victor C.M. Leung and Kun Yang, was essential for the success of the conference. We sincerely appreciate their constant support and guidance. It was also a great pleasure to work with such an excellent Organizing Committee; we are grateful for their hard work in organizing and supporting the conference. Particular thanks go to the Technical Program Committee, who completed the peer-review process for the technical papers and put together a high-quality technical program. We are also grateful to the Conference Manager, Natasha Onofrei, for her support and to all the authors who submitted their papers to the SmartGIFT 2020 conference.

We strongly believe that the SmartGIFT conference provides a good forum for all researchers, developers and practitioners to discuss all aspects of science and technology that are relevant to smart grids. We also expect that the future SmartGIFT conferences will be as successful and stimulating, as indicated by the contributions presented in this volume.

Maggie Cheng
Peng Yu
Yuan Hong
Huibin Jia

Conference Organization

Steering Committee

Imrich Chlamtac	University of Trento, Italy
Victor C. M. Leung	University of British Columbia, Canada
Kun Yang	University of Essex, UK

Organizing Committee

General Chair

Maggie Cheng	Illinois Institute of Technology, USA

General Co-chair

Shanchen Pang	China University of Petroleum, China

Technical Program Committee Chair and Co-chairs

Yuan Hong	Illinois Institute of Technology, USA
Peng Yu	Beijing University of Posts and Telecommunications, China
Huibin Jia	North China Electric Power University, China

Publicity and Social Media Chairs

Weizhi Meng	Technical University of Denmark, Denmark
Xuejun Li	Auckland University of Technology, New Zealand
Hongbin Sun	Changchun Institute of Technology, China

Publications Chairs

Yu-an Tan	Beijing Institute of Technology, China
Arun Kumar Sangaiah	Vellore Institute of Technology, India

Web Chairs

Ahmed Eissa	Middlesex University, UK
Yukun Dong	China University of Petroleum, China

Technical Program Committee

Baogang Li	North China Electric Power University, China
Bhalchandra Hardas	Ramdeobaba College of Engineering and Management, India

Fanqin Zhou	Beijing University of Posts and Telecommunications, China
Fei Zheng	Guilin University of Electronic Technology, China
Haibing Lu	Santa Clara University, USA
Hazim Jarrah	Auckland University of Technology, New Zealand
Hieu Nguyen	University of Quebec, Canada
Khan Ferdous Wahid	Airbus Group Innovations, Germany
Lei Shi	Institute of Technology Carlow, Ireland
Liansheng Tan	Central China Normal University, China
Lu Lu	Chinese Academy of Sciences, China
Mehboob Ul Amin	University Of Kashmir, India
Mohamed Riduan Abid	Al Akhawayn University, Morocco
Peng-Yong Kong	Khalifa University, United Arab Emirates
Peter Han Joo Chong	Auckland University of Technology, Australia
Pierluigi Siano	University of Salerno, Italy
Qingtao Zeng	Beijing Institute of Graphic Communication, China
Shangyu Xie	Illinois Institute of Technology, USA
Shunbo Lei	The University of Hong Kong
Stephan Cejka	Siemens Department of Corporate Technology, Austria
Sujie Shao	Beijing University of Posts and Telecommunications, China
Yinghui Ye	Xi'An University of Posts and Telecommunications, China
Yongjun Xu	Chongqing University, China
Zhixiong Chen	North China Electric Power University, China

Contents

Communications, Networks and Services

A New Similarity Measurement Method for the Power Load Curves
Analysis. 3
   Xin Ning, Ke Zhu, Yuanshi Deng, Rui Zhang, Qi Chen, and Zhong Li

Resource Prediction and Allocation Method for 5G C-RAN Based on
Power Internet of Things 14
   *Junfeng Lv, Zifan Li, Bozhong Li, Fang Chen, Zhengyuan Liu,
   and Peng Yu*

Design and Implementation of D2D Communication-Based 5G Cloud
Radio Access Networks Cell Outage Compensation Method. 27
   Junli Mao, Donghong Wei, Qicai Wang, Haoyu Wang, and Peng Yu

Service Capability Optimization Algorithm for Power Communication
Network Service Providers in Competitive Game Environment 42
   Zhi Li, Kai Duan, and Tingting Xu

Security and Stable Control

Multi-domain Cooperative Service Fault Diagnosis Algorithm
Under Network Slicing with Software Defined Networks 55
   Wei Li, Yong Dai, Yong Xu, Xilao Wu, Wei Li, and Peng Lin

Fault Diagnosis Algorithm Based on Service Characteristics Under
Software Defined Network Slicing 65
   Wei Li, Hao Cai, Chunxia Jiang, Ping Xia, Song Jiang, and Peng Lin

Security Situation Awareness and Interference Control Method for Power
Wireless Private Networks Based on Dynamic Baseline 77
   *Jin Huang, Weiwei Miao, Junzhong Yang, Xinglong Wang, Linshan Shi,
   Zhengyuan Liu, and Peng Yu*

Energy-Aware Blockchain Resource Allocation Algorithm with Deep
Reinforcement Learning for Trusted Authentication 93
   *Lifang Gao, Xiaotao Zhang, Tingfeng Liu, Huifeng Yang, Boxian Liao,
   and Jing Guo*

Internet of Power Things and Big Data

IPv6 Header Compression Scheme for Power Internet of Things.   107
   Wang Xiaoyu, Lu Xu, Liu Chuan, Tao Jing, and Liang Zhonghua

IPV6 Address Configuration Method in 6LoWPAN Oriented to the Internet
of Power Things.   117
   Lu Xu, Li Jianwei, Jiang Hao, Luo Dan, and Cao Han

A Resource Consumption Attack Identification Method Based
on Data Fusion.   128
   Libin Jiao, Yonghua Huo, Ningling Ge, Zhongdi Ge, and Yang Yang

Author Index   141

Communications, Networks and Services

A New Similarity Measurement Method for the Power Load Curves Analysis

Xin Ning[1], Ke Zhu[1], Yuanshi Deng[1], Rui Zhang[1], Qi Chen[2(⊠)], and Zhong Li[2]

[1] State Grid Sichuan Electric Power Research Institute, Chengdu 610072, Sichuan, China
[2] North China Electric Power University, Baoding 071003, Hebei, China

Abstract. In order to improve the quality of the power load curves similarity measurement, a new similarity measurement method based on Euclidean distance is proposed in this paper. Among the commonly used similarity measurement methods, Euclidean distance is not sensitive to the fluctuation of the load curves, which results in the lack of shape measurement capability. For the numerical distribution on the timeline is not concerned, the dynamic time warping (DTW) distance is not accord with the requirement of the power system load analysis. Focus on those issues, the proposed method introduced a correction factor that contains the dynamic characteristics of the numerical difference between two power load curves without compromising time warping. The advantages and performance of the proposed method are evaluated by similarity computing and clustering analysis. As shown in the experimental results of similarity computing, the proposed method performs as same as ED and DTW, but the calculating time is less than DTW. In the clustering analysis, it also decreases the calculating time from 3.9 s to 0.595 s compared with DTW and shows better clustering effect that make the Davies-Bouldin index from 0.438 for ED and 0.325 for DTW to 0.249.

Keywords: Power load curves · Similarity measurement · Euclidean distance · Cluster analysis

1 Introduction

A large number of valuable information contained in the power load curves has produced a great interest in power load analysis in recent years. The power load curves are always adopted to describe the variation of power load with time and reflect the characteristics and rules of users' electricity consumption. Generally, similarity measurement of the power load curves is popular to be used in load analysis [1]. As the basic of load forecasting and load pattern recognition [2], similarity measurement had various applications. Francisco et al. [3] used pattern sequence similarity for energy time series forecasting. Zhou and Li et al. [4] proposed a source-load-storage coordinated optimization model with source-load similarity. Singh [5] proposed a work on energy time series for users' behavioral analytics and energy consumption forecasting. Nagi, et al. [6] using support vector machines to detected nontechnical loss for metered customers in power utility.

M. Cheng et al. (Eds.): SmartGIFT 2020, LNICST 373, pp. 3–13, 2021.
https://doi.org/10.1007/978-3-030-73562-3_1

There are a lot of methods have been designed and proposed [7–14] to increase the quality of similarity measurement. Euclidean distance (ED) [7] and dynamic time warping distance (DTW) [8] are two typical examples. However, those two methods shown no applicable to varying degrees. Specifically, existing ED cannot recognize the change of curve shape. Besides, the calculating process of existing DTW made the timeline not aligned and had low efficiency, which is not suitable for the power system analysis. In addition to these two methods, Li and Yuan [9] considered the characteristic of vector difference and proposed a method for similarity estimate but only tested in graphic data set. Jia et al. [10] used imporved clustering method to evaluated the shape of the load curves. Yu et al. [11] presented a general guideline to find a better distance measure for similarity estimation based on statistical analysis of distribution models. Teeraratkul et al. proposed a shape-based approach to household electric load curves clustering and prediction based on DTW [12]. Although above methods [10–12] have achieved some results in load analysis, they all have complex calculation procedure that can't meet the requirement of load analysis for efficiency. Then a new similarity distance calculation method is urgently needed to be put forward.

This paper enumerates two classical distance-based similarity measurement methods, i.e. Euclidean distance and dynamic time warping distance, firstly. Then the comparison between the applicable scope, advantages and disadvantages are shown in detail. Aiming at the deficiency of similarity measurement and the characteristics of power load curves, this paper proposed a similarity measurement method, i.e. modified Euclidean distance (MED). Then the influence of the proposed improved algorithm in power load curve similarity measurement and its application effect in power load curve clustering are analyzed in the design experiment. The results are compared with Euclidean distance and dynamic time warping distance, respectively. The proposed MED method ensures accurate similarity calculation and increases the clustering results effectively.

The rest of the paper is organized in the following way. Existing ED method and dynamic time warping distance (DTW) are shown in Section II. In Section III, the proposed method, which is optimized by improvement factor is introduced with detailed analysis. Experimental validation of the proposed algorithm is given in Section IV to compare with existing ED and DTW. In Section V, conclusion is given to summarize the advantages of the proposed method.

2   Background

2.1   The Similarity of the Power Load Curves

For the power load curves, there are two factors influence the degree of similarity. The one factor is numerical similarity, which is reflected by the distance of two power load curves on the same sampling instant. The other factor, shape similarity, is the consistency degree of the dynamic changes of two load curves during the whole sampling time. As shown in Fig. 1, the distance between point a and point b reflect the power load curves numerical similarity. The shape similarity can be measured by comparing the shapes and variation trend of each curves.

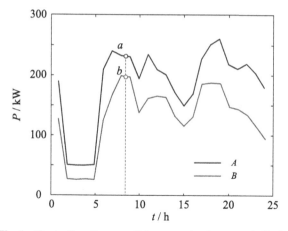

Fig. 1. Illustration diagram of the power load curves similarity.

2.2 Classic Similarity Measurement Methods

Among all the similarity measurement methods, Euclidean distance and dynamic time warping (DTW) distance are used commonly.

Euclidean Distance (ED). Suppose there are two load curves with length n, X ($X = x_1$, $x_2, \ldots, x_i, \ldots, x_n$) and Y ($Y = y_1, y_2, \ldots, y_i, \ldots, y_n$). To compare the similarity between X and Y, the Minkowski distance is used as:

$$D(X, Y) = \sqrt[r]{\sum_{i=1}^{n} (x_i - y_i)^r} \tag{1}$$

where r ($r \geq 1$) is a distance coefficient of Minkowski distance. Different values of r result in different similarity measurement methods. Euclidean distance is a special Minkowski distance when $r = 2$. So, the Euclidean distance between X and Y is:

$$D_{ED}(X, Y) = \sqrt{\sum_{i=1}^{n} (x_i - y_i)^2} \tag{2}$$

The process of Euclidean distance works is a strictly point-to-point calculation and simple to implement. But on account of using the square of the difference between x_i and y_i, Euclidean distance ignores the sign of ($x_i - y_i$). Therefore, Euclidean distance is not sensitive to the fluctuation of the load curves, which results in the lack of shape measurement capability.

Figure 2 shows three daily power load curves A, B and C of different users on the same date, where each curve has 24 sampling points (take samples once an hour). Calculating the Euclidean distance between A and B, A and C, the results are: D_{ED}(A, B) = 610.6907, D_{ED}(A, C) = 965.0453, D_{ED}(A, B) < D_{ED}(A, C).

The calculations results indicate that the load curve A is more similar with the load curve B. But load curve A is more similar with load curve C when comparing the shape and variation trend of three load curves as shown in Fig. 2.

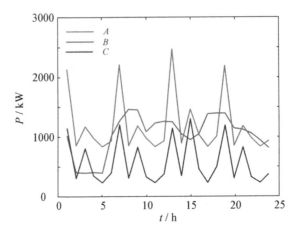

Fig. 2. Comparison of different load curves using Euclidean diatance.

Dynamic Time Warping (DTW) Distance. DTW based on the thinking of dynamic programming. This method searches for a shortest path through time stretching or warping, which makes the distance between two load curves is minimum. For time series: $X = \{x_1, x_2, ..., x_i\}$ and $Y = \{y_1, y_2, ..., y_j\}$. The definition of DTW is:

$$D_{DTW}(X, Y) = \begin{cases} 0, i = j = 0; \\ \infty, i = 0 \text{ or } j = 0; \\ D_b(x_1, y_1) + \min \begin{cases} D_{DTW}(Rest(X), Rest(Y)), \\ D_{DTW}(Rest(X), Y), \\ D_{DTW}(X, Rest(Y)) \end{cases} \end{cases} \tag{3}$$

where i and j are the length of the load curves X and Y respectively, $Rest(X) = \{x_2, ..., x_i\}$, $Rest(Y) = \{y_2, ..., y_j\}$. D_b is basic distance between x_i and y_j, usually be used in Euclidean distance.

Two points connected by a black line in Fig. 3 are the similarities of the load curves A and B. When several points correspond to one point, the time warping will occur. The sum of the distances of all the similarities is DTW distance.

In spite of DTW could recognize the load curves' shape features to some extent from Fig. 3, it's not a good choice for the power system load analysis. Because the timeline of the load curves is not fully aligned, DTW neglected the numerical distribution on the time line. This measure may cause over warping sometimes and not reflect the real change of the power load. In addition, DTW also have the shortcoming of high complexity.

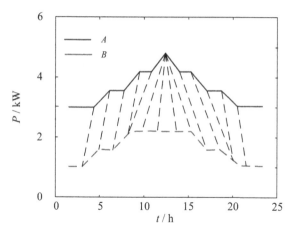

Fig. 3. Illustration diagram of DTW.

3 A New Similarity Measurement Method for the Power Load Curves

Generally, the similarity measurement method should be designed with specific problems. In the power system load analysis, both numerical similarity and shape similarity of the power load curves need to be concerned by similarity measurement method. Even though Euclidean distance and DTW are competitive for many fields, they have some shortcomings as a method to measure the similarity of the power load curves. Each of them could not consider numerical and shape similarity at the same time and the calculation of DTW has no advantage in speed.

Focus on these issues, a modified Euclidean distance (MED) for the power load curves is proposed as the new similarity measurement method. The MED between the load curves X ($X = x_1, x_2, ..., x_i, ..., x_n$) and Y ($Y = y_1, y_2, ..., y_i, ..., y_n$) is defined as:

$$D_{\text{MED}}(X, Y) = D_{\text{ED}}(X, Y) + \delta \tag{4}$$

where δ is a shape correction factor defined as:

$$\delta = \sqrt{\frac{1}{n}\sum_{i=1}^{n}(x_i - y_i - \mu)^2} \tag{5}$$

$$\mu = \frac{1}{n}\sum_{i=1}^{n}(x_i - y_i) \tag{6}$$

where $(x_i - y_i)$ is the difference between the corresponding point of two load curves on the same sampling instant.

The definition of MED shows that Euclidean distance can be made having shape measurement capability by introducing a correction factor and not compromising time warping. The correction factor δ is standard deviation of $(x_i - y_i)$, which represents the fluctuation degree of $(x_i - y_i)$. The more similar the two load curves are, the smaller the δ is. δ offsets the shortcoming of Euclidean distance that it couldn't consider the sign of $(x_i - y_i)$. The algorithm of MED is shown as follows, it can be easily implemented.

The algorithm of MED

Input: Daily load curves L_1 and L_2.
Output: Calculated distance.
Step 1: The mean value of the difference between L_1 and L_2 is calculated as (6).
Step 2: Calculating the distance of L_1 and L_2 by ED as ED (L_1, L_2).
Step 3: The parameter δ, which is the factor used to show the change trend of load curve is calculated as (5) ;
Step 4: Output the value combing ED (L_1, L_2) and Sig together as (4);

---

4   Experiment

To verify the feasibility and effectiveness of the proposed method, three described methods, i.e. ED, DTW and MED are applied to the load curves with similarity measurement and clustering. Then the comparisons between them are shown with detailed analysis. In this part, the experiment results of similarity measurement are shown at first. Then the clustering results of K-means are shown with analysis of Davies-Bouldin index (DBI), calculating time and iterations. Compile software: MatlabR2016a, operating system: Windows10, CPU: Inter(R) Core(TM) i3-8100, dominant frequency: 3.6GHz, internal storage: 16G, hard drive capacity:1T.

4.1   Experiment Data

Experiment data in this part is commercial and residential hourly load profiles for all TMY3 locations in the United States from Open EI [15]. This data set contains hourly load profile data for 16 commercial building types (based off the DOE commercial reference building models) and residential buildings (based off the Building America House Simulation Protocols). This data set also uses the residential energy consumption survey (RECS) for statistical references of building types by location.

4.2   Similarity Measurement

There are four load curves selected from the above data set adopted here to compare different similarity measurements, where the x-coordinate of the curves is the time point and the y-coordinate is the load value. The calculating results are shown in Table 1.

As shown in Table 1, the calculating results show that the calculation accuracy of MED is same as existing methods ED and DTW. Among all adopted methods, curve L_3 and L_4 are the closest, which conforms to the result of manual interpretation. And the second highest similarity is L_1 and L_4. For all methods, L_3 and L_4 owns the farthest distance. In terms of calculating time, as shown in Table 2, the calculating time of MED is little more than ED, but less than DTW (Fig. 4).

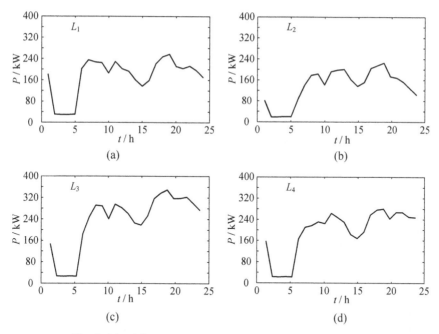

Fig. 4. The daily load curves selected for similarity calculation.

Table 1. The distance results for similarity calculation.

Distance	ED	DTW	MED
$D(L_1, L_2)$	221.4492	681.3785	253.6778
$D(L_1, L_3)$	301.1149	1140.8	340.2479
$D(L_1, L_4)$	193.2995	672.4120	222.0406
$D(L_2, L_3)$	445.5682	1600	490.4716
$D(L_2, L_4)$	342.0599	1275.8	379.8231
$D(L_3, L_4)$	**135.7468**	**427.7579**	**154.5377**

Table 2. The calculating time results for similarity calculation.

Methods	ED	DTW	MED
Time	**0.007456**	0.671990	0.007468

4.3 Clustering Analysis

K-means clustering method is applied in this part to analysis. Additionally, the K values adopted in different methods are all determined by ED for those methods. The reason for adopting it for three methods is to ensure the fairness of comparison. DBI is always used to confirm the value K. And the DBI is introduced as following.

A dispersion value S_i is defined in DBI to represent the dispersions of clusters i as:

$$S_i = \left\{ \frac{1}{T_i} \sum_{j=1}^{T_i} |X_j - A_i|^q \right\}^{\frac{1}{q}} \tag{7}$$

where T_i and A_i is the number of vectors and the centroid in cluster i, respectively. X_j represents the data point in clusters j. When q is taken as 1, it means the mean value of the distance from each point to the center. When q is taken as 2, it means the standard deviation of the distance from each point to the center. Both of them can be used to measure the degree of dispersion.

M_{ij} is the distance between vectors which are chosen as characteristic of cluster i and j and be calculated as:

$$M_{ij} = \left\{ \sum_{k=1}^{N} |a_{ki} - a_{kj}|^p \right\}^{\frac{1}{p}} \tag{8}$$

where a_{ki} is the k_{th} component of the n-dimensional vector a_i, which is the centroid of cluster i. Then R_{ij} is conducted to show the similarity between cluster i and j as:

$$R_{ij} = \frac{S_i + S_j}{M_{ij}} \tag{9}$$

where S_i and S_j are the dispersions of clusters i and j, respectively.

Then, DBI index is calculated as the mean value of R_i, which is the maximum value of R_{ij}.

$$\bar{R} = \frac{1}{N} \sum_{i=1}^{N} R_i \tag{10}$$

In this part, 640 daily load curves of 16 kinds commercial and residential customers of 4 different cities are chosen from adopted data set. As show in Fig. 5, K is better chosen 5 to realize clustering. Then the clustering results of three different similarity measurements are produced.

Clustering Results. The clustering results of three similarity measurement methods are shown in Fig. 6, 7 and 8, respectively, where (a) is the clustering results and (b) is the calculated clustering center.

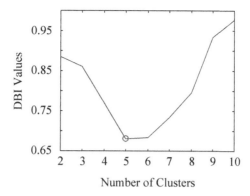

Fig. 5. DBI index during various clusters number.

As shown in Fig. 6, 7 and 8, all three methods realize clustering efficiently. But the clustering results are a little bit different. The clustering results are shown in Table 3, MED decreases the DBI index compared with ED and DTW, which results in a more accurate clustering result. Besides, the clustering time of MED is the shortest, while the calculating time of DTW is much longer.

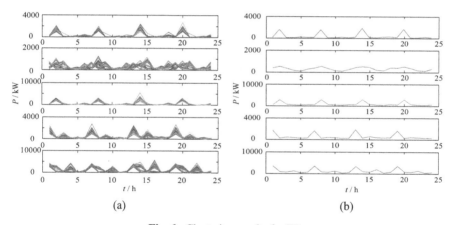

Fig. 6. Clustering results for ED.

4.4 Analysis of Experimental Results

It can be seen that the proposed MED method realized the similarity measurement effectively. And the proposed MED method having the same calculation result with ED, the calculating time of the proposed MED method is between the time of ED and DTW. It can be seen from the clustering results that the MED method increases the clustering results, which results in a lower DBI index. With the decrease of DBI, the proposed algorithm increases the iterations compared with ED. However, the iterations and calculating time of MED are all less than DTW.

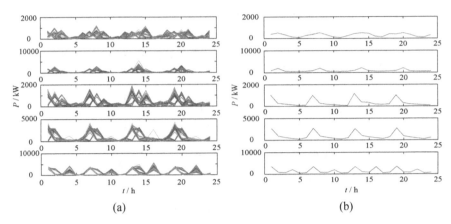

(a)   (b)

Fig. 7. Clustering results for DTW.

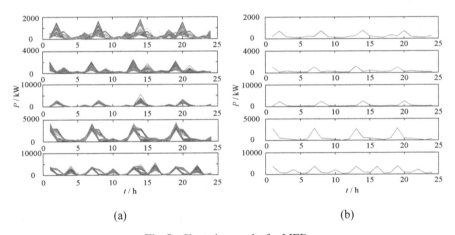

(a)   (b)

Fig. 8. Clustering results for MED.

Table 3. The clustering results for three methods.

	DBI	Calculating time	Iterations
ED	*0.4382*	0.637700 s	**14**
DTW	0.3248	*3.902797 s*	29
MED	**0.2487**	**0.595371 s**	23

5   Conclusion

A modified similarity measurement method based on conventional Euclidean distance is designed and proposed in this paper. Focus on the issues for existing similarity measurement methods, proposed method reflected the change of curve shape without complex calculation. Besides, the advantages of proposed are proved by similarity calculation and clustering analysis in experimental validation. Compared with Euclidean distance and dynamic time warping distance methods, the proposed method increases the clustering effect without increasing the calculating burden significantly. In future, proposed algorithm will be implied into other clustering algorithms and the parameter that reflects the change in shape will be improved to increase the clustering effect further.

Acknowledgments. This work is supported by the State Grid Corporation of China (52199719002M).

References

1. Shi, L., Zhou, R., et al.: New energy-load characteristic index based on time series similarity measurement. Electr. Power Autom. Equipment **39**(5), 75–81 (2019)
2. Lin, R., Wu, B., Su, Y.: An adaptive weighted parson similarity measurement method for load curve clustering. Energies **11**, 2466 (2018)
3. Alvarez, F.M., Troncoso, A., Riquelme, J.C., Ruiz, J.S.A.: Energy time series forecasting based on pattern sequence similarity. IEEE Trans. Knowl. Data Eng. **23**(8), 1230–1243 (2011). https://doi.org/10.1109/TKDE.2010.227
4. Zhou, R., et al.: Source-load-storage coordinated optimization model with source-load similarity and curve volatility constraints. Proc. CSEE **40**(13), 4092–4101 (2020)
5. Singh, S., Yassine, A.: Big data mining of energy time series for behavioral analytics and energy consumption forecasting. Energies **11**, 452 (2018)
6. Nagi, J., Yap, K.S., Tiong, S.K., et al.: Nontechnical loss detection for metered customers in power utility using support vector machines. IEEE Trans. Power Delivery **25**(2), 1162–1171 (2010)
7. Yu, K., Guo, G., et al.: Quantum algorithms for similarity measurement based on euclidean distance. Int. J. Theor. Phys. **59**, 3134–3144 (2020)
8. Mei, J., Liu, M., Wang, Y., Gao, H.: Learning a mahalanobis distance-based dynamic time warping measure for multivariate time deries vlassification. IEEE Trans. Cybern. **46**(6), 1363–1374 (2016)
9. Li, Z., Yuan, J.: An estimation similarity measure method based on the characteristic of vector difference. Int. J. Inf. **14**(3), 1067–1074 (2011)
10. Jia, H.M., He, G.Y., Fang, C.X., Li, K.W., Yao, Y.Z., Huang, M.M.: Load forecasting by multi-hierarchy clustering combining hierarchy clustering with approaching algorithm in two directions. Power Syst. Technol **31**, 33–36 (2007)
11. Yu, J., Amores, J., Sebe, N., Radeva, P., Tian, Q.: Distance learning for similarity estimation. IEEE Trans. Pattern Anal. Mach. Intell. **29**, 451–462 (2008)
12. Teeraratkul, T., O'Neill, D., Lall, S.: Shape-based approach to household electric load curve clustering and prediction. IEEE Trans. Smart Grid **9**(5), 5196–5206 (2018)
13. Gao, M., Gong, T., Lin, R., et al.: A power load clustering method based on limited DTW algorithm. In: Information Technology, Networking, Electronic and Automation Control Conference, Chengdu, pp. 253–256. IEEE (2019)
14. Lin, R., Wu, B., Su, Y.: An adaptive weighted pearson similarity measurement method for load curve clustering. Energies **11**(9), 2466 (2018)
15. https://openei.org/datasets/dataset/commercial-and-residential-hourly-load-profiles-for-all-tmy3-locations-in-the-united-states

Resource Prediction and Allocation Method for 5G C-RAN Based on Power Internet of Things

Junfeng Lv[1], Zifan Li[1], Bozhong Li[1], Fang Chen[1], Zhengyuan Liu[2], and Peng Yu[2(✉)]

[1] State Grid Information and Telecommunication Branch, Beijing 100761, China
[2] Beijing University of Posts and Telecommunications, Beijing 100876, China
yupeng@bupt.edu.cn

Abstract. The construction of power Internet of things is an important practice of network power strategy, which can greatly improve the coordination and data connectivity between various businesses of power grid. With the continuous maturity of 5G technology, it also can be used as an alternative access network solution. Facing the power Internet of things, this paper discusses the feasibility of C-RAN cloud wireless access network architecture in the power Internet of things access. Furthermore, this paper designs the base station traffic prediction algorithm based on LSTM and the network resource allocation algorithm based on genetic algorithm, which improves the utilization efficiency of network resources, and is of great significance for the future access network in the power industry to save construction costs and energy consumption.

Keywords: Ubiquitous power internet of things · Electric power wireless network · C-RAN · Genetic algorithm

1 Introduction

On May 22, 2020, the third session of the 13th National People's Congress opened in Beijing. Premier Li Keqiang made a government work report. The report pointed out that in the current epidemic situation, we should continue to "promote the reduction of production costs of enterprises". The policy of reducing the electricity price of industry and Commerce by 5% will be extended to the end of this year ". It is estimated that the electricity tariff will be reduced by 92.6 billion yuan in the whole year [1, 2]. While firmly implementing the decision-making and deployment of the Party Central Committee and the State Council, the State Grid actively saves resources and opens up current, increases investment in power grid by 9.9%, reduces costs and continuously improves input-output efficiency [3–5].

Investment in the construction of the power Internet of things can effectively promote the recovery of the upstream and downstream of the industrial chain, and play an important role of power grid investment. After the completion of the power Internet of things, it will greatly improve the connectivity and coordination of data between the various

M. Cheng et al. (Eds.): SmartGIFT 2020, LNICST 373, pp. 14–26, 2021.
https://doi.org/10.1007/978-3-030-73562-3_2

services of the power grid, which is conducive to the implementation of the requirements of cost reduction and efficiency improvement. From the concept, the power Internet of things is to use the most cutting-edge mobile Internet communication, artificial intelligence and other modern technologies, aiming at each link and process of the power system, so that the heterogeneous terminal equipment in each link of the power system can be connected with each other, and the working state and communication situation of all terminal equipment in the system and the system itself can be comprehensively monitored and controlled through the perception layer. The efficient use of system information realizes the intelligent driving of the system through data, which greatly improves the utilization efficiency of resources. At present, about 450 million devices of various types are connected to the power grid system, including electricity meters and various types of equipment used for information collection, system protection and equipment control. According to the latest planning of the State Grid, it is estimated that there will be about 2 billion heterogeneous terminal devices in the system by 2030. At that time, the power Internet of things will become the biggest Internet of things with the largest number of terminal devices State circle.

In order to maintain an IoT ecosystem with such a large number of access devices, it is very important to save the construction cost and energy consumption of the power Internet of things, and the access network is the largest part of the construction cost and energy consumption of the system, so the selection of access network solution is very important for the whole system. At present, there are three kinds of access network schemes, which are 2G, 3G, 4G and 5G technologies of mobile public network, 230MHz and 1800MHz technologies of power wireless private network, and lpwa low-power WAN technologies, including Lora and NB-IoT technologies. Different access network schemes have different advantages and disadvantages, which form complementary to each other to a certain extent. With the continuous development of 5G wireless network, C-RAN, as a network architecture that can provide high band and wide access, has the characteristics of wide coverage and high bandwidth, and can be widely used in switching stations, distribution rooms, charging piles, power consumption information collection and other services [6].

The first section of this paper mainly introduces the characteristics of C-RAN access network architecture and the current research focus of the architecture, and expounds the applicability of the architecture for the power Internet of things. In the second section, aiming at the resource allocation problem of the power Internet of things based on C-RAN, this paper first proposes an access site traffic prediction algorithm based on LSTM model, and analyzes the prediction performance of the model through the predicted data. In the third section, based on the predicted traffic results, this paper proposes a network resource allocation algorithm based on genetic algorithm, and makes a comparison with existing algorithms in many aspects. The simulation results show that the performance of the algorithm is better than the existing algorithm. In the fourth section, the two algorithms proposed in this paper are summarized respectively.

2   Introduction of C-RAN Access Network Architecture

The C-RAN architecture is an improvement on the distributed base station architecture proposed by China Mobile Research Institute in 2010. The core idea is to divide the

BBU and RRH which need to be fixed into two independent parts and place them separately [7]. The two are connected through optical fiber. All baseband processing units (BBBs) are uniformly placed in the BBU pool [8], sharing baseband processing resources and corresponding supporting facilities, such as air conditioning and other refrigeration equipment, reduce the construction cost and energy consumption of supporting facilities. The architecture has four main advantages, namely, reducing energy consumption, saving cost, easy to upgrade and expand, and greatly improving the utilization efficiency of baseband processing resources (Fig. 1).

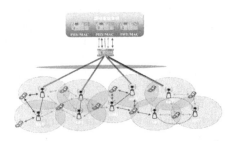

Fig. 1. C-RAN network architecture diagram.

C-RAN architecture has great advantages in network adaptability, energy saving and base station construction cost saving. In order to give full play to the advantages of the architecture, there are some problems to be solved, the most important of which is to achieve the prediction of base station traffic and the reasonable allocation of network resources. At present, the research work on C-RAN energy saving mainly focuses on the energy saving of RRH and transmission network. The energy-saving research of BBU is still in its infancy. How to intelligently allocate network resources to save energy consumption of BBU pool while maintaining good network performance has high research value.

The network resource allocation model discussed in this paper adopts 5G access network solution, and specifically uses C-RAN as access network architecture. The focus of the research is to allocate network resources intelligently through algorithm. In the power Internet of things, C-RAN network structure can be used to carry high-definition video monitoring, mobile operation and other large bandwidth services, as well as remote control business, feeder automation and other control services, and online power assets physical ID and other large connection services.

3   Traffic Prediction Algorithm of Access Station Based on LSTM Model

3.1   The Change Law of Node Station Flow

In the area covered by power 5G C-RAN, the location and communication status of some dynamic equipment such as intelligent inspection equipment are generally in continuous change, which also leads to a relatively regular, temporal or spatial change of base station load. There are two kinds of changes in time: large span oscillation and small span oscillation. The large span oscillation is often caused by the mobile changes of dynamic equipment, which can lead to a regular change of the base station load, and the change range is also relatively large, which is usually in hours. However, the small span oscillation is often caused by the access and disconnection of the service terminal equipment or the change of the communication status of the equipment. This kind of vibration is random and difficult to predict, and the change range is small. The time span is usually in minutes [9]. Large span oscillation and small span oscillation constitute the change curve of base station service load.

Taking the data of public network communication network as an example, as shown in Fig. 2, the load change of a station for several consecutive days is shown. It is obvious that the traffic shows regular fluctuations, and each traffic waveform is similar. Figure 3 shows the daily flow fluctuation of the station, with two obvious peaks. At the same time, due to the existence of small span oscillation, the curve has strong suddenness.

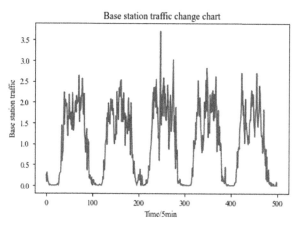

Fig. 2. Multi day load changes of base stations.

3.2 LSTM Traffic Prediction Model and Training Result Analysis

After preprocessing the original data, the preprocessed data is used to train the LSTM traffic prediction model. The MAE is used as the loss function. The function is shown in Eq. (1), which can better reflect the actual situation of the predicted value error.

$$MAE = \frac{\sum_{i=1}^{n} |y_i - x_i|}{n} \tag{1}$$

Figure 4 shows the learning curve of the LSTM traffic prediction model. When the model is trained to 20 epochs, the loss function MAE of the training set and the verification set is about 0.1, and the model tends to converge. After training, the loss

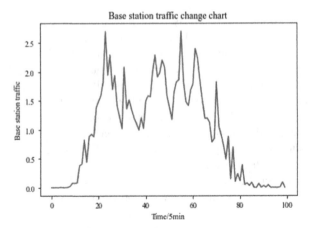

Fig. 3. Single day load change of base station.

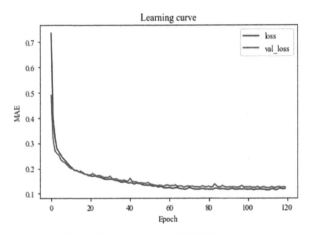

Fig. 4. Learning curve of LSTM model.

function value of the verification set is about 0.06, and the training situation of the model is good.

Figure 5 shows the comparison chart of the prediction results of LSTM traffic prediction model. By using the LSTM traffic prediction model after training, the traffic prediction data within a certain period of time can be obtained. By comparing the traffic prediction data obtained with the actual data, it can be found that the fitting degree of the two curves is good, and the prediction curve can better reflect the change of the flow. At the same time, it also can be found that the traffic prediction data in a certain period of time can be obtained. Although the prediction curve can also reflect the actual situation of traffic to a certain extent, there is still a certain deviation between the two curves, which also shows that the short-term irregular small span oscillation of network traffic is difficult to learn through the model. To sum up, the model has a strong learning ability for the regular large-span oscillation, but it is difficult to predict the irregular

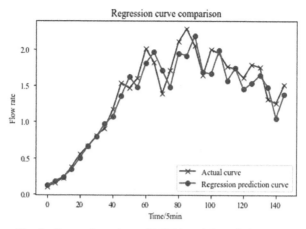

Fig. 5. Comparison chart of LSTM model prediction data.

small span oscillation, but it can also better reflect the flow changes, and the model has a high accuracy of flow prediction.

4 Network Resource Allocation Algorithm Based on Genetic Algorithm

4.1 Optimization Objective of C-RAN Model

After obtaining the characteristics of different types of traffic in the network, considering the diversity of Internet of things equipment terminals, it provides different power IoT terminals with network quality services, which can maintain network QoS and reduce energy consumption to the greatest extent.

To reduce the energy consumption of access network, we need to rely on the scheduling algorithm of BBU pool to realize the effective utilization of BBU, so as to reduce the overall energy consumption of BBU pool. The energy consumption of BBU pool mainly considers two parts, the operation energy consumption of BBU and the migration energy consumption of tasks in RRH in BBU. The operation energy consumption includes the energy consumption of calculation resources and the energy consumption of supporting facilities. Because the energy consumption of BBU in sleep state is far lower than that in operation, the overall energy consumption of BBU pool can be greatly reduced by shutting down the BBU with low load and constraining the remaining BBU load in a reasonable range [10]. In order to achieve the above purpose, it is necessary to migrate the services in the BBU, and the task migration will also generate a certain amount of energy consumption, that is, task migration energy consumption [11]. The task here can be regarded as one-to-one correspondence with the Internet of things services on the wireless side. Generally, the energy consumption of task migration is directly proportional to the amount of tasks. Now assume that all BBUs in the BBU pool are the same, and there is no difference in baseband processing capacity, downlink network rate and rated power.

The mathematical model of the system is as follows (2)–(10):
Objective function:

$$Min \, P_{total}(t) = \alpha_{total}(t) + \beta_{total}(t) + P_{static}(t) \tag{2}$$

Cost function:

$$\alpha_{total}(t) = \sum_{i=1}^{n} Pb_i(t) \tag{3}$$

$$Pb_i(t) = \gamma \cdot \sum_{z=1}^{h} G_i^z(t) \cdot S_z(t) + Pb_{basic}(t) \tag{4}$$

$$\beta_{total}(t) = \delta \cdot \sum_{z=1}^{h} G_{i,j}^{'z}(t) \cdot S_z'(t) \tag{5}$$

$$\varphi(\mu_i^m) = \alpha \cdot \mu_i^m \tag{6}$$

Environmental constraints:

$$\gamma, \delta, \alpha > 0 \tag{7}$$

$$\sum_{m=1}^{M} \varphi(\mu_i^m) \leq P_{max} \tag{8}$$

$$\sum_{m=1}^{M} c_i^m \leq C_{Max} \tag{9}$$

$$\mu_i^m = c_i^m \tag{10}$$

Equation (2) is the expression of the objective function, $P_{total}(t)$ is the sum of the energy consumption of the system. All we have to do is to reduce the value of the objective function as much as possible while maintaining QoS. $\alpha_{total}(t)$ is the total energy consumption of BBU operation, mainly the baseband data processing energy consumption; $\beta_{total}(t)$ is the total energy consumption of data migration; $P_{static}(t)$ is the total static energy consumption of the system, such as air conditioning refrigeration energy consumption and forward link energy consumption.

Equation (3) is the expression of the total energy consumption of BBU operation, n is the total number of BBUs in the BBU pool, and $Pb_i(t)$ is the energy consumption of the ith BBU in time t.

Equation (4) is the calculation formula of $Pb_i(t)$, h is the total number of tasks, $G_i^z(t) = 1$ is a boolean variable, indicating that at time t, the z-th task is assigned to the i-th BBU, $S_z(t)$ is the task amount of the z-th task, γ is the correction weight of the running energy consumption, $Pb_{basic}(t)$ is the basic energy consumption when the BBU is turned on. This part of energy consumption is fixed, and this part of energy consumption will not be generated until the BBU is closed.

Equation (5) is the expression of the total energy consumption of data migration. $G_{i,j}^{'z}(t)$ is a boolean variable, $G_{i,j}^{'z}(t) = 1$ indicates that the z-th task migrates from the i-th BBU to the j-th BBU at time t, $S_z'(t)$ represents the data volume of the task during data migration, and δ is the correction weight of the task energy consumption.

Equation (6) is the power expression of the corresponding task, μ_i^m is the processing rate of the m-th task in the i-th BBU. According to the relevant references, the instantaneous power of the task has a linear relationship with the corresponding processing rate, α is the correction weight of the power.

Equation (7) indicates that γ, δ and α are all greater than 0.

Equation (8) is the constraint on the instantaneous power of the BBU, which means that in the i-th BBU, the total power consumed by all tasks shall not be greater than the rated maximum power of the BBU, M is the total number of tasks at this time of the BBU, and P_{max} is the rated maximum power of the BBU.

Equation (9) is the constraint on the downlink rate of the BBU, which means that the sum of the downlink rates of all tasks in the i-th BBU is not greater than the rated maximum downlink rate of the BBU, c_i^m is the downlink network rate of the M task in the i-th BBU, and C_{Max} is the rated maximum value of the downlink network rate of the BBU.

Equation (10) is the constraint on the processing speed and downlink network speed of the m-th task in the i-th BBU. These two values should be roughly equal, that is $\mu_i^m = c_i^m$, to meet the user's quality of service when receiving downlink data [12].

The above problem is a typical NP hard problem. In this paper, the genetic algorithm of intelligent optimization algorithm is selected to solve the problem.

4.2 Design of Network Resource Allocation Algorithm Based on Genetic Algorithms

Genetic algorithm (GA) is a method of searching for the optimal solution by simulating the process of biological evolution, which makes the population continuously select, cross and mutate. It is often used to optimize and solve NP hard problems [13].

Coding: RRH can be continuously migrated in different BBUs in the C-RAN architecture, so that the load of the BBU is always in a reasonable range, and the overall energy consumption of the BBU pool can be saved by turning off some of the lower load BBUs. The first step of the genetic algorithm is to code the chromosomes. Assuming the number of RRH is m and the number of BBU is n in the access network system, there are M genes in the chromosome. Each gene has n different choices. Genes at different locations represent different RRH. The number of genes indicates the BBU to which the RRH will migrate. Each chromosome is a RRH migration scheme. After coding, a certain number of chromosomes are generated, and the genes in the chromosomes are randomly generated. These chromosomes will begin to evolve as the first generation population.

Selection: By calculating the energy consumption of baseband processing and task migration, the total energy consumption required for each chromosome, is the fitness of the model, can be obtained. The higher the total energy consumption, the lower the environmental adaptability. Using the championship selection method as the selection strategy, two chromosomes were randomly selected from the population at a time to compare the total energy consumption of the two chromosomes [14]. The chromosome with lower total energy consumption was selected to enter the offspring population, while the chromosome with higher total energy consumption was eliminated and repeated until each individual in the population was selected.

Crossing: Chromosome crossing is the exchange of genes between two chromosomes, in which a more adaptable chromosome may be obtained and the crossed chromosome will be added to the offspring population.

Mutation: After chromosome crossover, there is a certain probability that a gene mutation will occur. At this time, two numbers are randomly generated to determine the specific location of the mutation and the value of the mutation. The mutation operation can avoid the population convergence to the local optimal solution, maintain the diversity of population genes to a certain extent, and ensure the algorithm to fully search for the optimal solution.

The flow chart of the algorithm is shown in Fig. 6 below.

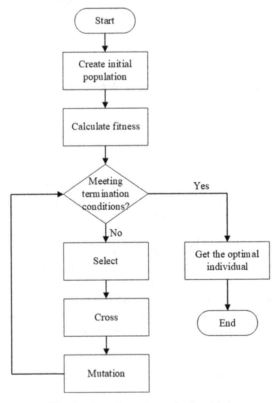

Fig. 6. Flow chart of genetic algorithm.

4.3 Simulation Results

Business types and corresponding security requirements and single-point bandwidth in the power Internet of Things are shown in Table 1 below [15]:

Table 1. Power business scenarios

Business category	Safety requirements	Single point bandwidth
Protection	Extremely high	2000 Kbit/s
Control	Extremely high	20 Kbit/s
Information detection	Relatively high	10 Kbit/s
Video	Relatively high	2000 Kbit/s
Marketing	High	1000 Kbit/s

The system parameters used in the simulation experiment are shown in Table 2:

Table 2. C-RAN system parameter setting table

Parameter name	Numerical value
Number of BBU pools	1
Number of BBUs	30
Number of RRHs	60
BBU running power	50 w
Maximum rated power of BBU	500 w
Maximum rated downlink rate of BBU	50 Mb/s

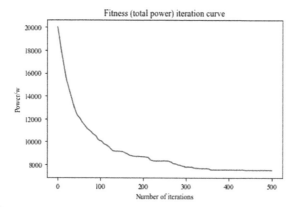

Fig. 7. Genetic algorithm fitness iteration graph.

As shown in Fig. 7, for the fitness iteration diagram of the genetic algorithm, it can be seen that the average fitness of the population has converged and the algorithm has a better effect on the total energy consumption when iterating to 400 generations.

As shown in Fig. 8, the genetic algorithm-based network resource allocation algorithm and the existing algorithm proposed for this paper have the same trend in the number of active BBUs in the system under the same network environment. However, the number of active BBUs in the genetic algorithm is slightly more than the existing algorithm.

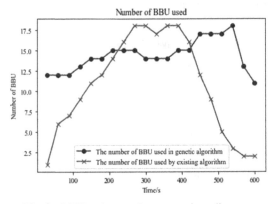

Fig. 8. BBU active number comparison diagram.

As shown in Fig. 9, the task migration times of the two algorithms are compared. The task migration times of the genetic algorithm are always less and the curve is more stable, which will not cause large network fluctuations. The task migration times of existing algorithms are more and the curve changes more.

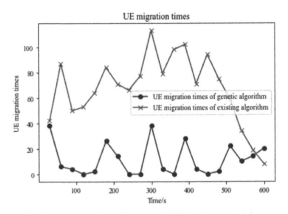

Fig. 9. Comparison diagram of UE migration times.

The number of active BBUs of the genetic algorithm is slightly more than that of the existing algorithms, so the genetic algorithm consumes slightly more energy in the

BBU baseband processing than the existing algorithms, but the number of task migration is much lower than that of the existing algorithms, so the energy consumption of task migration is much lower than that of the existing algorithms. As shown in Fig. 10, the total power consumption of the two algorithms is compared. The total power consumption of the genetic algorithm is lower than that of the existing algorithms, and the curve of the total power consumption is more stable. This shows that the genetic algorithm is better than the existing algorithms in the comprehensive performance.

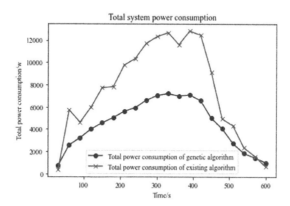

Fig. 10. Comparison of total power consumption of system.

5   Conclusion

For the access network construction of the power Internet of Things, this paper analyses the feasibility of 5G C-RAN network hosting, and proposes LSTM-based traffic prediction algorithm and genetic algorithm-based network resource allocation algorithm respectively. LSTM-based traffic prediction algorithm has good prediction accuracy and can accurately reflect the trend of traffic change. Compared with the existing algorithms, the genetic algorithm-based network resource allocation algorithm has fewer task migrations and lower overall system energy consumption. It can better balance the energy consumption of BBU baseband processing and task migration, and allocate network resources reasonably. This framework and method can provide theoretical basis for the construction and resource allocation of the power Internet of Things. Next, we need to propose a more intelligent and self-adapting dynamic resource adjustment method for different business QoS needs in order to further improve the flexibility of the network and ensure the quality of service of the power Internet of Things.

References

1. https://www.sgcc.com.cn/html/sgcc_main/col2017021449/2020-05/21/202005211710551 62880954_1.shtml

2. https://www.sgcc.com.cn/html/sgcc_main/col2017021449/2020-05/23/202005231804279
45966122_1.shtml

3. https://www.sgcc.com.cn/html/sgcc_main/col2017021449/2020-05/20/202005201020567
02377177_1.shtml

4. https://www.sgcc.com.cn/html/sgcc_main/col2017021449/2020-05/22/202005220934316
07370914_1.shtml

5. https://www.sgcc.com.cn/html/sgcc_main/col2017021449/2020-05/18/202005181929529
03559108_1.shtml

6. He, E., Geng, P., Zhang, H.: Networking scheme of the power wireless private network based on edge computing. Telecommun. Sci. (Z2), 248–255 (2019)

7. Chunli, Y.E.: BBU-RRH Efficient Load Balancing Analysis for C-RAN. Beijing University of Posts and Telecommunications, Beijing (2016)

8. Weng, B., Wang, J., Liu, L.: The analysis of C-RAN network application. Design. Tech. Posts Telecommun. (11), 57–60 (2012)

9. Chen, W.: Research on load forecasting method and migration scheduing strategy for carrier migration in C-RAN. Xidian University, Xi'an (2017)

10. Guo, H.: Research on Energy Saving Algorithm in C-RAN Architecture. Beijing University of Posts and Telecommunications, Beijing (2017)

11. Chien, W.C., Lai, C.F., Chao, H.C.: Dynamic resource prediction and allocation in C-RAN with edge artificial intelligence. IEEE Trans. Ind. Inform. **15**, 4306–4314 (2019)

12. K, W.: Research on C-RAN based Energy-Efficient Resource Allocation Problems. University of Science and Technology of China, Hefei (2017)

13. H, Z.: Research on Robust Scheduling of Manufacturing Enterprice under MTO Mode. Chongqing University of Technology, Chongqing (2015)

14. Huang, Z.: Comparing for Algorithms on Researching of Active Power Optimization in the Power Systems. Nanchang University, Nanchang (2013)

15. Chen, B., Wu, Q., Lai, M.: Applied research of TD-LTE power wireless broadband private network. Power Syst. Commun. (11), 82–87 (2012)

Design and Implementation of D2D Communication-Based 5G Cloud Radio Access Networks Cell Outage Compensation Method

Junli Mao[1], Donghong Wei[1], Qicai Wang[1], Haoyu Wang[2], and Peng Yu[2(✉)]

[1] The 54Th Research Institute of CETC, Shijiazhuang, Hebei, China
[2] Beijing University of Posts and Telecommunications, Beijing 100876, P.R. China
yupeng@bupt.edu.cn

Abstract. As 5G officially enters the commercial era, heterogeneous networks will face massive data traffic requirements and frequent communication overheads in the future. How to ensure user service quality and reduce energy consumption is a key challenge for major operators. The wireless cloud network is considered as one of the architectures to solve this problem, and its biggest advantage is that it can quickly and flexibly access the dynamically configured resources of the shared pool. Existing research on cell interruption compensation technology mainly focuses on the optimization and adjustment of parameters of neighboring cells, which easily affects the original network topology. This paper proposes a cell interruption compensation scheme based on D2D technology, focusing on routing and resource allocation. In two stages of research, a routing algorithm based on the attributes of social networks was proposed, and the encounter model in social networks was introduced in the resource allocation link. Simulation results show that the scheme further reduces system energy consumption and delays the growth rate of energy consumption compared with traditional methods.

Keywords: Power cyber-physical system · Routing re-construction · Genetic algorithm

1 Introduction

Cloud-Radio Access Networks (C-RAN) is a new type of broadband access network based on distributed remote base stations to centralize all or part of the baseband processing resources, form a baseband resource pool, and uniformly manage and dynamically allocate them. In the process of large-scale coverage of 5G networks, the quality of user services is reduced due to the massive data traffic requirements of heterogeneous dynamic users and frequent communication overhead. C-RAN is considered to be a solution to this problem. The architecture has so far received extensive attention from academia and industry [1].

Due to the complex 5G Cloud-Radio Access network environment, the probability of node failure is greatly increased. For any network operator, mitigating the impact

M. Cheng et al. (Eds.): SmartGIFT 2020, LNICST 373, pp. 27–41, 2021.
https://doi.org/10.1007/978-3-030-73562-3_3

of different potential failures on users is a very important task. If a problem occurs at a certain node, the efficiency and speed of dealing with the problem manually cannot meet the user's service quality requirements. That is, when the network fails, it must be guaranteed that the line can have a transmission rate of hundreds or tens of megabytes per second and can independently handle related issues. The specific embodiment of this kind of network autonomous management is called Self-Organizing Network (SON), which is defined as a network that can dynamically adapt to network changes and has the ability to optimize its performance and solve its failures. In SON, it mainly includes three function: self-configuration, self-optimization and autonomy. This article mainly focuses on the research of Cell Outage Compensation (COC) in autonomous healing, introduces Device-to-Device (D2D) technology in the terminal compensation link, and builds an energy-efficient collaborative compensation optimization model to solve the network optimization of edge users The compensation problem can solve the service interruption problem of users at the edge of the network while taking into account energy consumption.

In the Sect. 1, this article briefly introduces the research background and significance of the article, explains the importance of using SON in the 5G C-RAN network, and proposes the advanced nature of introducing D2D technology in the SON autonomous healing process; the Sect. 2 introduces the current Research progress on self-organizing networks and D2D technology, and analyzes some problems in the current research; Sect. 3 builds a system model with the goal of optimal system energy consumption in D2D paired connection, and divides the entire system model into routing The two stages of selection and resource allocation are to build an optimization model for research, and propose solutions for the optimization model; Sect. 4 verifies the correctness and advancement of the above optimization model through system simulation, and compares the results with the relay selection based on energy efficiency The schemes are compared; Sect. 5 is the conclusion of the full text, summarizes the work completed by the full text, and analyzes the shortcomings in the content of the article, pointing out the direction for the next step of research.

2   5G C-RAN Network, Autonomous Healing Technology and D2D Technology Research Status

Because C-RAN has the above advantages of Centralization, Cloud, Cooperation and Clean. A large number of related studies have also been carried out on C-RAN at home and abroad. Literature [2, 3] studied the application of SDN (Software Defined Network) and NFV (Network Function Virtualization) technology in C-RAN; literature [4, 5] studied the deployment of C-RAN architecture Research on strategy; and literature [6] research on resource allocation in C-RAN network and so on. But for the dense network like 5G C-RAN, there is little research on its autonomy.

For the study of COC in autonomous healing, the existing method of achieving COC is to optimize the interruption area by adjusting the antenna gain or transmission power of the neighboring base station. In literature [7], a method for adjusting the transmission power and antenna is studied The comprehensive optimized cell interruption compensation mechanism of inclination is mainly to achieve the purpose of optimization compensation by changing the relevant parameters of adjacent base stations, but it is easy to cause changes in the network topology and the neighboring relationship of the base stations, and may introduce additional interference and energy consumption In the patent [8], the invention proposes a cell interruption compensation scheme that performs cooperative compensation through neighboring cells. The scheme is mainly to detect the channel service status of the neighboring base station and select the best idle channel through a game algorithm. The idle channel can provide network services for the interrupted cell users. However, in this scheme, the distribution of users is relatively regular, without considering the complicated characteristics of the distribution of users in the 5G network, and the re-allocation of the idle channels of neighboring cells will affect the network quality to a certain extent. The above two methods have certain limitations in the 5G C-RAN network composed of dense sites.

Due to its many advantages, D2D technology has also become a key research object at home and abroad in recent years. In terms of routing, literature [9] proposed a routing algorithm based on the user's actual location and a routing algorithm based on channel quality. On this basis, literature [10] re-optimized the system using bit error rate as the standard. Reference [11] uses channel utilization as a reference standard to optimize D2D routing strategies. In terms of resource allocation, the traditional research idea is to transform the resource allocation problem into a control variable optimization model, and then solve it through a heuristic algorithm or an optimization algorithm. Reference [12] introduces a resource optimization algorithm for the D2D communication system with the highest energy efficiency. This paper simplifies the resource allocation problem into two processes: power control and spectrum selection. The power control algorithm and channel selection algorithm based on KM are applied. Reference [13] uses network coding technology in D2D communication, which improves the quality and range of D2D communication while reducing the load of the base station, making the overall throughput of the network better. Reference [14] introduces access control, power control and spectrum selection into the resource allocation process, and uses the weighted bipartite graph maximum matching algorithm to solve, which also greatly improves the overall performance of the network.

The above research has contributed to the practical application of D2D technology, but there are still many shortcomings. In actual situations, the user's geographic location will often change, leading to inaccurate positioning, resulting in a great deal of traditional routing strategies limitation. Moreover, the above methods are mainly aimed at static networks. In a complex and rapidly changing 5G network environment, each change in the network environment will bring a huge amount of recalculation, and it is difficult to meet the highly dynamic network environment. This article will introduce the relevant attributes of social networks on the basis of the original research, and combine each terminal device with the person it uses. The social relationship of people is relatively fixed within a certain period of time. The introduction of social attributes makes the terminal device better. It adapts to the rapidly changing network environment, finds the best paired device, and reduces the additional energy consumption of the system.

3   System Model and Optimization Model

3.1   System Model

The schematic diagram of the D2D interrupt compensation scenario is shown in Fig. 1. The three RRHs cover areas A, B, and C, where the RRH of area A fails due to external factors, and some terminals on the edge of the network need to be terminals in areas B and C. For D2D compensation, there are three types of users: ordinary users, D2D directly connected users and D2D single relay connected users under the coverage of RRH.

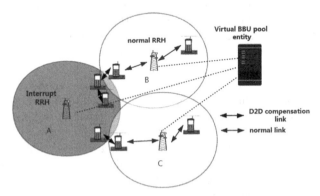

Fig. 1. Schematic diagram of D2D-based user compensation in 5G C-RAN (Color figure online)

In Fig. 1, one of the RRHs fails for some reason and requires the network to quickly repair itself to achieve the goal of maintaining network performance. For network edge users, D2D compensation can be performed by other terminal devices under the coverage of adjacent RRHs to restore the network. The black arrow in the figure represents the normal link between the terminal device and the RRH, and the blue arrow between the terminal devices indicates D2D Compensate the link. In order to facilitate research in the network topology, the relevant network environment is defined as follows:

Let the interrupted RRH in the area be detected, the set of RRH is B = {b_1, b_2,..., b_i,..., b_N}, The interrupted user set is U_0 = {u_1, u_2, ..., u_j,..., u_M}, The active user set is VA = {v_1, v_2, ..., v_k,..., v_L}. X = {x_{jk}} represents the selection relationship matrix of relays between users, 1 means connected, 0 means not connected; $X' = \{x'_{ij}\}$ indicates the association between active users and RRH, 1 means connected, 0 means not connected. In order to facilitate processing, each interrupted user is connected to the network through appropriate D2D selection. The distance between RRH and active users is set to D_{ij}, The distance between active users and interrupted users is D_{jk}, The path loss index is γ, The transmit power of RRH is P_i^R, The transmit power of active users is P_j^V, Then when x_{jk} = 1 and x'_{ij} = 1, D2D link is officially established. The channel between the D2D links is a Rayleigh fading channel. By default, mutual interference between D2D pairs and ordinary users is not considered. When allocating resources to D2D to users, adopt the method of orthogonal shared channels. The signal-to-noise ratio of the two-hop link from the source terminal to the relay terminal and the relay terminal to the destination terminal is as follows:

$$SNR_{ij} = \frac{|h|^2}{N_0} D_{ij}^{-\gamma} P_i^R \tag{1}$$

$$SNR_{jk} = \frac{|h|^2}{N_0} D_{jk}^{-\gamma} P_j^V \tag{2}$$

N_0 is the noise spectral density, γ is the path loss index, h represents the channel coefficient, and $|h|^2$ follows an exponential distribution with a mean of 1. For compensated user j, when connected to active user k, the corresponding channel capacity can be simply set as:

$$C_m = \min\{\log(1 + SNR_{ij}), \; \log(1 + SNR_{jk})\} \tag{3}$$

For the probability of successful connection between two hop links, which is expressed as:

$$p_1 = p(SNR_{ij} \geq \theta) = \exp(-\frac{N_0 \theta \, D^\gamma}{P_i^R}) \tag{4}$$

$$p_2 = p(SNR_{jk} \geq \theta) = \exp(-\frac{N_0 \theta \, D^\gamma}{P_j^V}) \tag{5}$$

θ is the signal to noise ratio threshold.

During the connection of the D2D link, the channel time slot occupied by the link is also an important consideration. We assume that the time slot occupied by the interrupted user i and the active user k is t_{jk}, $0 < t_{jk} < 1$. Based on the above, it can be concluded that the energy efficiency *EE* of the entire system can be expressed as:

$$EEm = \frac{\sum\limits_{k=1}^{L} \sum\limits_{j=1}^{M} C_{jk}}{\sum\limits_{k=1}^{L} \sum\limits_{j=1}^{M} \sum\limits_{i=1}^{N} (1+\alpha)(t_{jk}P_i^R + (1-t_{jk})P_j^V) + (1+t_{jk})P_{cr} + P_{ct}} \tag{6}$$

α is a constant, which is generally determined by the power and the energy conversion efficiency when the power is changed. P_{cr} and P_{ct} is energy consumption of source terminal and destination terminal, which also set as constant. We can dynamically adjust the three values of t_{jk}, P_i^R and P_j^V to change the value of *EE*. We assume that the transmission power of each source terminal is constant, so the transmission power of the link is determined by the link connection x_{jk} and x_{ij}', and our optimization goal for each link is to obtain the maximum energy efficiency of the system.

$$\max_{\{t_{jk}\},\{x_{jk}\},\{x_{ij}'\}} EE \tag{7}$$

Constraints included:

$$\forall j, k, c_{jk} \geq c_M \tag{8}$$

$$\forall j, k, \sum_{j=1}^{M} \sum_{k=1}^{L} x_{jk} t_{jk} \leq 1 \tag{9}$$

$$\forall i, j, p_{ij} \geq p^* \tag{10}$$

$$\forall i, k, p'_{jk} \geq p^* \tag{11}$$

$$\forall i, j, SNR_{ij} \geq \theta \tag{12}$$

$$\forall j, k, SNR_{jk} \geq \theta \tag{13}$$

$$P^R_{\min} < \forall i, P^R_i \leq P^R_{\max} \tag{14}$$

$$P^V_{\min} < \forall j, P^V_j \leq P^V_{\max} \tag{15}$$

C_m is the channel capacity threshold for user information transmission, and p^* is the threshold for link connection success rate. $P^R_{\max}$ and $P^R_{\min}$ are the minimum and maximum limit values of RRU transmit power respectively, $P^V_{\max}$ and $P^V_{\min}$ are the minimum and maximum limit values of active user transmit power, respectively.

3.2 Optimization Model of Routing Process

Generally speaking, increasing the transmission power of the source terminal can improve the success rate of link connection, but with the increase of the transmission power, it will also bring greater interference, reduce the signal-to-noise ratio, and make it difficult to guarantee the communication quality. It brings more waste of resources and cannot be achieved in actual scenarios. To determine the appropriate transmit power to make the signal-to-noise ratio and link connection success rate meet the conditions, the optimization goal of this process is to find the minimum connection energy consumption, and then determine the appropriate Transmit power

$$Q1 = \min_{\min}(P^R_i + P^V_j + P_{cr} + P_{ct}) \tag{16}$$

Constraints satisfy Eqs. 8–15.

In recent years, people have paid more and more attention to the analysis and research of social networks. Many behavior patterns at the user level often affect the mutual perception between devices, and related research on social networks has slowly penetrated into other fields. D2D terminals are generally light and easy to carry. Today's smart terminals already have the ability to perceive the effective information in the user, the surrounding environment and the network they are in [15]. This article introduces social network attributes into the routing process. The schematic diagram is shown in Fig. 2:

In social network theory, the three key parameters that measure the intimacy between two users are centrality, similarity, and trust. Therefore, based on these three quantities,

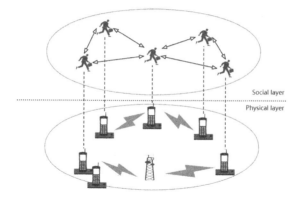

Fig. 2. D2D model with social attributes

an abstract modeling is performed to synthesize a utility function and set a threshold. When the intimacy between users is greater than this threshold, it indicates that the success rate of D2D communication between devices is also high. The social layer uses this scheme to filter the set of relay devices that meet the conditions, and then uses the existing scheme to select relay terminals at the physical layer, which can effectively avoid unnecessary energy loss and reduce the probability of communication failure [16].

Centrality. Degree centrality indicates the importance of a node, which shows the popularity of a person in the concept of a social network. The higher the popularity, the more frequently he communicates with others, and the easier it is to establish connections with others. In the physical layer, the centrality of the device is determined by the number of times the probe message is received. Here we set the number of times the active user v_j receives the probe message as $Freq(v_j)$, then the formula for the centrality is:

$$Cen(v_j) = Freq(v_j) \Big/ \sum_{n}^{N} Freq(v'_n) \tag{17}$$

N represents the total number of nodes in the network, v' represents other active users in the network.

Similarity. The current research on the route discovery process mainly focuses on parameters such as channel quality and transmission rate, and ignores the user's communication needs. As a result, devices that meet the conditions during the route selection process refuse to communicate because of insufficient intimacy, resulting in a waste of resources. In fact, in social networks, due to the relatively fixed social circle, people are always willing to exchange information and share resources with users who have similar social behaviors or hobbies. The possibility of people with similar odors becoming friends Big. In the routing process, if D2D has the same interest in candidate users and source users, the success rate of communication will increase. In the concept of social networks, the similarity between two users is expressed by the cosine of their social distance, and the formula is:

$$Sim(u_i, v_j) = |I(u_i) \cap I(v_j)| \Big/ \sqrt{|I(u_i)| * |I(v_j)|} \tag{18}$$

I represents the user's interest set.

Trust. At the social level, the closer the two people are, the more trust they have in each other. Corresponding to the physical layer, the more interactions between the two devices, the greater the trust between each other. Social relations are related to the frequency and duration of contact between two people. The communication frequency and communication duration between the two devices can be obtained from the cloud BBU pool. Here, we abstract the trust between the two users u_i and u_j as a formula:

$$Tru(u_i, v_j) = \frac{CF(u_i, v_j)}{\sum_{t \in N} CF(u_i, v_t)} * \frac{CT(u_i, v_j)}{\sum_{t \in N} CT(u_i, v_t)} \tag{19}$$

$CF(ui, vj)$ represents the communication frequency between two users. $CT(ui, vj)$ represents the communication time between two users.

Utility Model. Based on the basic concepts in the above three social networks, this paper summarizes the utility function as:

$$U(u_i, v_j) = \alpha \cdot Tru(u_i, v_j) + \beta \cdot Sim(u_i, v_j) + \gamma \cdot Cen(u_i) \tag{20}$$

Among them, the three values of α, β, γ are parameters, satisfying $\alpha + \beta + \gamma = 1$, which can be changed according to the actual situation. In this formula, set the threshold value U_{thr}, and make the utility function value of the surrounding nodes and the destination node greater than. The threshold nodes are put into the candidate relay set to complete the preliminary screening.

3.3   Optimization Model of Resource Allocation Process

Within the coverage of an RRU, we can first analyze the geographic distribution of each device. If the distance between the two users meets the distance for D2D communication, the two terminal devices have the conditions to establish D2D communication. The introduction of the encounter model in resource allocation can allocate channel resources according to the communication success rate to avoid waste. One of the important quantities in the encounter model is the duration of the encounter. In [17], a general gamma distribution used to describe the encounter time of two terminals is proposed $\Gamma(k, u)$, where k and u are two variables It is related to the mean and variance. In literature [18], the probability of successful communication between the two devices can be calculated according to the meeting time of the two devices. The formula is:

$$\omega_{jk} = 1 - \int_0^{T \min} f(x, k, u)dx \tag{21}$$

Where T_{min} represents the shortest encounter time that two terminals can establish D2D communication, $f(x, k, u)$ is the probability density function of the duration of their encounter, both of these values are provided by the base station.

After obtaining the success rate of D2D communication between two devices, we can set the priority according to this success rate to allocate resources. The greater the communication success rate P, the higher the allocation priority β, the definition of β is as follows:

$$\beta = k(\omega_{jk} - a) + 1 \tag{22}$$

α represents the average value of the D2D communication success rate in the current range, which is determined according to the specific conditions of each area, and can be given by the BBU pool, k represents a parameter, and the method of taking the parameter k is described below. Since the priority β is derived from the communication success rate P, the two values should have a proportional relationship. At this time, we can get the following formula:

$$\frac{\beta 1}{\omega 1} = \frac{\beta 2}{\omega 2} = \frac{\beta 3}{\omega 3} = \ldots = \frac{\beta n}{\omega n} \tag{23}$$

Combining formula (22) and formula (23), we can find the value of k: $k = \frac{1}{a}$, so for the allocation priority β, the formula is:

$$\beta = (\frac{1}{a})(\omega_{jk} - a) + 1 \tag{24}$$

After obtaining the priority β, we can sort the connected D2D terminal pairs and allocate resources. Channels with high channel quality are preferentially assigned to users with high priority.

4 Case Analysis

In order to be able to verify the effectiveness of the algorithm proposed in this paper, we analyze it through simulation examples. Simulate and verify the whole in two stages.

4.1 Routing Stage

The D2D model used in the simulation is a single-relay two-hop model. It is assumed that there are two RRU-covered cell areas. One of the RRUs is faulty and is set to area A. The coverage of the RRUs without faults is set to area B. The coverage radius of a single RRU is 200 m, assuming that the number of terminals to be compensated in area A is 10, the number of active interruptions in area B is 20, and the terminal positions in each area follow an even distribution. The channels established by each terminal node are Rayleigh channels, and noise interference between terminals is not considered, only uplink communication is considered, and the working mode of each node is a half-duplex mode. Set the three parameters of the utility function to $\alpha = \beta = \gamma = \frac{1}{3}$, indicating that the three values of trust, centrality, and similarity share the same proportion. First,

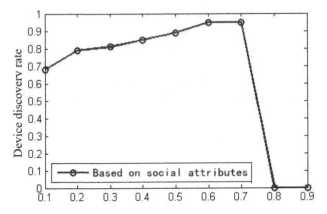

Fig. 3. The threshold value of the utility function

we determine the threshold of the utility function according to the simulation, as shown in Fig. 3:

With the increase of the preset threshold, the communication success rate of the scheme slowly increases to 95%, at this time the threshold is 0.6 or 0.7. After that, it quickly decreased to 0. The reason is that as the threshold increases, candidate terminals that do not meet the conditions are gradually eliminated, and the remaining terminals in the candidate list are more and more in line with the destination terminal, so the success rate of communication is also higher and higher. However, if we continue to increase the threshold, it indicates that the similarity between the two devices is getting higher and higher. At this time, the terminal that meets the conditions may no longer exist. Therefore, as shown in the figure, we set the threshold of the utility function to 0.7. After that, we compare this solution with OEE routing scheme [19] and random routing scheme. The path loss index and the distance from the source terminal to the destination terminal are used as independent variables. The performance comparison is shown in Fig. 4 and Fig. 5:

From the above simulation results, it can be seen that as the path loss index and the distance between the two terminals continue to increase, the energy consumed by the three schemes is increasing. Under a specific path loss index, it can be seen that the energy consumption of the OEE routing scheme is less than the random routing scheme, and the routing scheme based on social attributes needs to go further on the basis of the OEE routing scheme to achieve the predetermined goal.

4.2 Resource Allocation Stage

In the process of resource allocation, D2D will reuse the channel resources of cellular users. We assume that there are three RRU coverage areas in a scenario, where RRU1 is temporarily unable to provide services due to failures. It is set to use channel resources under RRU coverage. There are 6 groups of users, including 3 ordinary users and 3 pairs of D2D users. The link between D2D users has been successfully established. In the

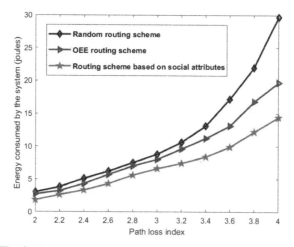

Fig. 4. Comparison of energy consumption of the three schemes

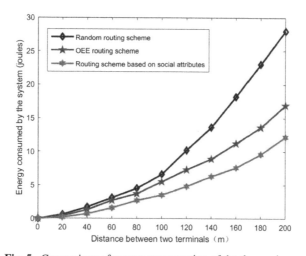

Fig. 5. Comparison of energy consumption of the three schemes

network, there are 6 available channels that are orthogonally allocated to each group of users. The simulation scene diagram is shown in Fig. 6:

According to the configuration of the system simulation parameters, the average communication rate of ordinary users and D2D-to-users in the current environment is 427kbit/s and 438kbit/s, respectively, and it is assumed that the channels of the two communication users in different time periods The rates meet the Rayleigh distribution with mean values of 427 kbit/s and 438 kbit/s respectively . The overall rate of the system is obtained by Eq. 25, where V_{al} is the overall rate of the system, $V(u)$ is the channel

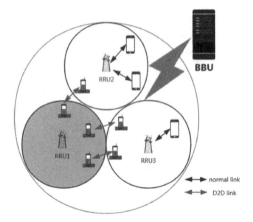

Fig. 6. Simulation scene diagram

transmission rate of ordinary users), $V(d, d)$ is the channel transmission rate of D2D users.

$$Val = \sum_{k=1}^{3} V(u_k) + \sum_{i=1}^{3} \sum_{j=1}^{3} V(u_i, v_j)\beta_i \tag{24}$$

We then set up two communication scenarios. The communication success rates of the three sets of D2D pairs are $P_1 = 0.3$ $P_2 = 0.5$ $P_3 = 0.7$. At this time, the average communication success rate is $\alpha = (P_1+P_2+P_3)/3 = 0.5$, then their resource allocation priorities are $\beta_1 = 0.6$, $\beta_2 = 1.0$, $\beta_3 = 1.4$. The success rates of the three sets of D2D pairs in scene 2 are $P_1 = 0.1$ $P_2 = 0.5$ $P_3 = 0.9$, then the average communication success rate is $\alpha = (P_1 + P_2 + P_3)/3 = 0.5$, then their resource allocation priorities are $\beta_1 = 0.2$, $\beta_2 = 1.0$, $\beta_3 = 1.8$. The total transmission rate of the system in two scenarios is shown in Fig. 7 and Fig. 8:

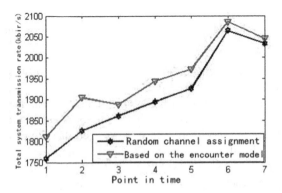

Fig.7. The total system transmission rate of the two schemes in scene one

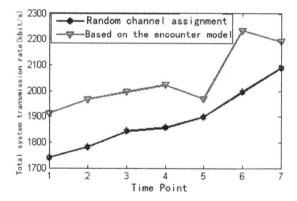

Fig. 8. The total system transmission rate of the two schemes in scene two

Through Fig. 7 and Fig. 8 we can conclude that the resource allocation method using the encounter model will increase the overall transmission rate of the system. However, in scenario two, the increase in system rate is more obvious, because in scenario two, the distribution variance of the success rate of the D2D pair connection is larger, so that a better channel is allocated to a better D2D time band The greater the revenue, the better performance.

5 Conclusion

This paper mainly studies the scheme of using D2D technology for fast network compensation for the edge area of the network covered by RRU after the 5G C-RAN network is interrupted. Different from the traditional scheme of changing the base station inclination or network parameters for optimization and compensation, fully taking into account the portable characteristics of D2D devices, introducing social network attributes, adding social attributes and encounter models to the two D2D connection processes of routing and resource allocation can fully avoid The problem of unnecessary network loss due to user selfishness is of great significance for reducing the waste of system power resources and channel resources.

However, there are still some imperfect work in this article, which only considers the D2D connection method of single-hop unicast. In reality, a D2D terminal with a good geographic or social location may undertake the transmission of multiple pairs of D2D pairs. Devices with relatively remote locations may require multi-hop D2D pairs to complete network compensation. Moreover, this article ignores the mutual interference of various D2Ds in communication. In a complex network, signal interference is one of the problems that must be solved.

The next work will focus on solving the signal interference problem in the D2D connection process, and at the same time investigate and explore the network topology more realistically, and explore the difficulties and possibilities of multi-hop multicast D2D technology. At the same time, since the D2D technology has a relatively small range of network compensation, in the case of a large-scale network failure, it is also the direction of efforts to explore more autonomous solutions with lower energy consumption and better compensation effects.

References

1. Aqeeli, E., Moubayed, A., et al.: Power-aware optimized RRH to BBU allocation in C-RAN. IEEE Trans. Wirel. Commun. **17**(2), 1311–1322 (2018)
2. Orsino, A., Araniti, G., Wang, L., et al.: Enhanced C-RAN architecture supporting SDN and NFV functionalities for D2D communications, pp. 3–12 (2016)
3. Akyildiz, I.F., Lin, S.C., Wang, P.: Wireless software-defined networks (W-SDNs) and network function virtualization (NFV) for 5G cellular systems: an overview and qualitative evaluation. Comput. Netw. **93**, 66–79 (2015)
4. Huang, X.,Wang, W., Zhao, P.: C-RAN architecture deployment scheme for 5G scale evolution. Post Telecommun. Des. Technol. **11**, 39–43 (2019)
5. Tao, Y., Song, H.: Discussion on construction strategy of transmission network for 5G C-RAN. Post Telecommun. Des. Technol. **515**(1), 86–91 (2019)
6. Abdelnasser, A., Hossain, E.: Resource allocation for an OFDMA cloud-RAN of small cells overlaying a macrocell. IEEE Trans. Mob. Comput. **15**(11), 2837–2850 (2015)
7. Wenjing, L., Yin Mengjun, Y., Peng, , Y., et al.: Cell interruption compensation mechanism based on joint optimization of power and inclination. J. Electron. Inf. Technol. **05**, 195–201 (2015)
8. Hui,T., Ping, Z., Shaoshuai, F.: A cooperative compensation service method for solving the service interruption in residential areas: China, 201510388363.1. 25 November 2015
9. Sreng, V., Yanikomeroglu, H., Falconer, D.D.: Relayer selection strategies in cellular networks with peer-to-peer relaying. In: Vehicular Technology Conference, 2003. VTC 2003-Fall. 2003 IEEE 58th, vol. 3, pp. 1949 –1953. IEEE (2003)
10. Wang, C.L., Syue, S.J.: A geographic-based approach to relay selection for wireless ad hoc relay networks. In: VTC-Spring 2009, pp. 1–5. IEEE (2009)
11. Zhu, X., Wen, S., Wang, C., et al.: A cross-layer study: information correlation based scheduling scheme for device-to-device radio underlaying cellular networks. In: 2012 19th International Conference on Telecommunications (ICT), pp. 1–6. IEEE (2012)
12. Wei, S., Huang, X., Wang, W.: Interference coordination and resource optimization of D2D communication system based on energy efficiency. Comput. Appl. Res. (07), 206–209. 2018321
13. Zhao, Z., Shao, S., Sun, J.: The research of network coding in D2D communication system. In: International Conference on Computer, Networks and Communication Engineering (ICCNCE 2013), pp. 396–400. Atlantis Press (2013)
14. Feng, D.: Research on wireless resource allocation of D2D Communication. University of Electronic Science and technology (2015)
15. Mao, H., Feng, W., Ge, N.: Social-aware cooperation among mobile terminals for wireless downlink transmission. Commun. China **12**(9), 1–10 (2015)
16. Stratogiannis, D.G., Tsiropoulos, G.I., Cottis, P.G.: Bandwidth allocation in wireless networks employing social distance aware utility functions. In: IEEE Globecom Workshops, pp. 543–548. IEEE (2011)
17. Guo, J., Liu, F., Zhu, Z.: Estimate the call duration distribution parameters in GSM system based on K-L divergence method. In: International Conference on Wireless Communications, Networking & Mobile Computing, pp. 2988–2991. IEEE (2007)

18. Li, W.: Research on D2D communication terminal discovery and resource allocation algorithm based on social characteristics (2016)
19. Amin, O., Lampe, L.: Opportunistic energy efficient cooperative communication. Wirel. Commun. Lett. IEEE 1(5), 412–415 (2012)

Service Capability Optimization Algorithm for Power Communication Network Service Providers in Competitive Game Environment

Zhi Li[✉], Kai Duan, and Tingting Xu

Marketing Service Center (Measurement Center) of State Grid Chongqing Electric Power Company, Chongqing, China

Abstract. In order to solve the problem of low resource utilization rate of multiple power communication network service providers, this paper proposes a power communication network resource management system, which consisting of the self-built power communication network service provider, the third-party power communication network service provider, a power communication network resource allocation center, and the demand side of power communication network. Secondly, the competitor's service cost coefficient is solved to obtain the competitor's competitive strategy, using the predictive mechanism of the service cost coefficient probability density function; and the reaction function based inference process is transformed to obtain the Jacobin iterative formula of service capability. Finally, a service capability optimization algorithm based on Jacobi iteration is proposed. In the simulation experiment part, the competition game model is simulated, which proves that the algorithm is more in line with the real environment than the competition game under the complete information. It is more practical for the power company to choose the power communication network service provider.

Keywords: Power communication network · Service provider · Resource allocation · Game

1 Introduction

In the power sector, with the rapid development of smart grid services, the demand for power communication networks is increasing [1]. The network infrastructure built by the power company is gradually unable to meet the development needs of the smart grid business. In the network-as-a-service environment, the network serves as a service, and the network builder provides services to the society. With the development of technologies such as network virtualization and network management, network as a service has been gradually applied to various fields [2]. In this context, the power company can adopt the self-built power communication network to transmit power services according to business needs, and can also lease third-party communication network resources to

M. Cheng et al. (Eds.): SmartGIFT 2020, LNICST 373, pp. 42–51, 2021.
https://doi.org/10.1007/978-3-030-73562-3_4

meet the needs of power communication services. Based on this, the service providers that provide the power communication network are divided into Self-built power communication network Service Providers (SSP), Third-party power communication network Service Provider (TSP).

Aimed at the problem of low resource utilization caused by the dynamics and complexity of the network environment, a network weight training algorithm based on the gradient descent algorithm is proposed [3]. The literature [4], for the QoS constraint problem of power communication service, a game model consisting of resource allocation center, resource demand side and resource provider is proposed, and a resource allocation mechanism that can achieve the balance of interests of all parties in the game. Literature [5] models the use of network resources at various times, and analyzes the resource utilization of different time periods. Based on this, a Bayesian-based virtual network resource allocation strategy is proposed to model the network resource allocation problem. The resource allocation problem for time and energy perception significantly improves resource utilization. Aiming at the problem of excessive network energy consumption, an energy-saving transmission strategy for cached wireless networks is proposed [6]. Aiming at the problem of low revenue for network resource providers, a joint optimization algorithm for resource allocation under heterogeneous networks is proposed [7]. Aiming at the problem of network resource waste, a task-oriented service quality resource demand forecasting method is proposed [8].

From the analysis of existing research results, it is known that power companies urgently need to obtain network resources through various channels to meet the development of smart grid services. Network as a service has become a trend. However, how electric power companies choose to build their own power communication networks and rent third-party communication networks has become an urgent problem to be solved. In order to solve this problem, this paper first constructs a game model of power communication network service provider in network as a service environment. Secondly, it analyzes how to realize the resource allocation algorithm of power company's revenue maximization under incomplete information. Finally, the rationality of the model is verified by simulation experiments.

2   Model Design

The power communication network resource management architecture is shown in Fig. 1. It includes SSPs, TSPs, power communication network resource allocation centers, and demand side of power communication network (DS). DSs can make resource requests to the power communication network resource allocation center. The power distribution network resource allocation center provides DS with network resources with certain network service capabilities according to resource requirements and resource status of SSPs and TSPs.

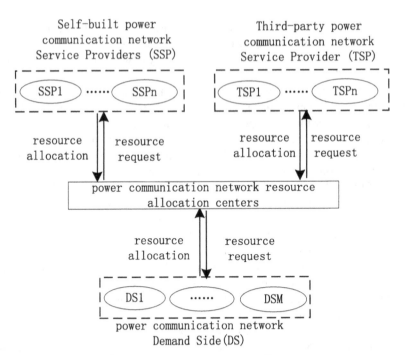

Fig. 1. Resource management architecture

The power communication network resource management architecture is shown in Fig. 1. It includes SSPs, TSPs, power communication network resource allocation centers, and demand side of power communication network (DS). DSs can make resource requests to the power communication network resource allocation center. The power distribution network resource allocation center provides DS with network resources with certain network service capabilities according to resource requirements and resource status of SSPs and TSPs.

In terms of resource demand, the resource demand indicated by the mth power service to the resource allocation center is used, and based on this, the total resource demand submitted by the m power services to the resource allocation center is $\Lambda = \sum_{m=1}^{M} \lambda_m$.

In the provision of power communication network services, it is considered that the purpose of providing services by the power communication service providers is to obtain profits. The difference between a cooperative game and a non-cooperative game is whether there is a binding agreement between the interacting parties. If there is, it is a cooperative game; if not, it is a non-cooperative game. The game theory that economists talk about generally refers to non-cooperative games. Since cooperative game theory is more complicated than non-cooperative game theory, its theoretical maturity is far less than non-cooperative game theory. Therefore, this paper uses a non-cooperative game model to describe the competitive relationship between power communication network service providers. The way to compete is generally to increase the quality of network services by increasing investment, or to attract more customers by lowering prices. Use indicates the total, quantity, where the quantity is expressed in terms of the quantity used $r - m$.

In different time periods, the operation of *SSPi* and *TSPj* needs to invest different funds. In this paper, $c_k^{s,i}$, $c_k^{t,j}$ is used to represent the operating cost of s at time k, The calculation is performed using Eq. (1), where k_1^s, k_1^t represent input factors for customer service and human resources, k_2^s, k_2^t represent input coefficients for network construction and operation. Use $u_k^{s,i}$ and $u_k^{t,j}$ to represent the network data processing capacity of *SSPi* and *TSPj*. The larger the processing capacity, the better the performance of the network, and the constraint condition is $0 < u_k^{s,i} < 1, 0 < u_k^{t,j} < 1$. Use q to represent *SSPi*, *TSPj* for the receiving rate of the service request, the higher the receiving rate, the better the network quality of *SSPi*, *TSPj*. Since the third-party network service provider needs to rent the interface of the power company communication network, it can be integrated with the power communication network service. Therefore, the relationship between k_2^s, k_2^t is $k_2^t > k_2^s$.

$$\begin{cases} c_k^{s,i} = k_1^s u_k^{s,i} + k_2^s q \\ c_k^{t,j} = k_1^t u_k^{t,j} + k_2^t q \end{cases} \tag{1}$$

In the resource allocation center, resource allocation is performed according to the resource demand quantity of the power communication network, the price of the service provider, and the quality of service. The final determined resource allocation ratios for each service provider are calculated using Eq. (2). Where $f_k^{s,i}$ and $f_k^{t,j}$ are used to represent the market proportion of *SSPi* and *TSPj*, and $f_k^{s,i} + f_k^{t,j} \le \Lambda$. Use a_i and a_j^* indicate the relationship between the service quality of *SSPi*, *TSPj* and $f_k^{s,i}, f_k^{t,j}$. Use b_{ij}, b_{ji}^* to indicate the probability that *SSPi*, *TSPj* can substitute each other.

$$\begin{cases} f_k^{s,i} = a_i u_k^{s,i} - \sum_{l=1,l\neq i}^{m} u_k^{s,l} - \sum_{j=1}^{r-m} b_{ij} u_k^{t,j} \\ f_k^{t,j} = a_j^* u_k^{t,j} - \sum_{h=1,h\neq j}^{r-m} u_k^{t,h} - \sum_{i=1}^{m} b_{ji}^* u_k^{s,i} \end{cases} \tag{2}$$

Based on the above analysis, $U_k^{s,i}$ and $U_k^{t,j}$ are used to represent the effects of *SSPi*, *TSPj*, and the formula (3) is used for calculation. Use $\pi_k^{s,i}$ and $\pi_k^{t,j}$ to represent the profit of *SSPi* and *TSPj*, and use formula (4) to calculate.

$$\begin{cases} U_k^{s,i} = P^s - c_k^{s,i} \\ U_k^{t,j} = P^t - c_k^{t,j} \end{cases} \tag{3}$$

$$\begin{cases} \pi_k^{s,i} = f_k^{s,i} \cdot U_k^{s,i} \\ \pi_k^{t,j} = f_k^{t,j} \cdot U_k^{t,j} \end{cases} \tag{4}$$

3   Competitive Game Objective Function

In the current market economy environment, the service price of the power communication network is relatively consistent and open. Therefore, the competitive approach of

SSPi and TSPj is to improve the quality of service as much as possible. However, improving the quality of service requires more human and material resources, which affects the cost of services and thus the value of k_1^s and k_1^t. In order to win in the competition, k_1^s and k_1^t of SSPi and TSPj will be kept as confidential as possible. In the process of competition, SSPi and TSPj will estimate their competitor's competitive strategy by collecting historical data of competitors. In this paper, the prediction mechanism of the probability density function based on the service cost coefficient is used to solve the competitor's service cost coefficient, so as to obtain the competition strategy of the competitor.

Use $f(k_1^s)$ to represent the probability density function of k_1^s for SSPi, and k_1^s has a value range of (a, b) (where, $0 < a < b$), so $\int_a^b f(k_1^s) d(k_1^s) = 1$. Use $f(k_1^t)$ to represent the probability density function of k_1^t for TSPj, and k_1^t has a value range of (c, d) (where, $0 < c < d$), so, $\int_c^d f(k_1^t) d(k_1^t) = 1$. So, the expected values of k_1^s and k_1^t are calculated by formula (5).

$$\begin{cases} E(k_1^s) = \int_a^b k_1^s \cdot f(k_1^s) d(k_1^s) \\ E(k_1^t) = \int_c^d k_1^t \cdot f(k_1^t) d(k_1^t) \end{cases} \tag{5}$$

Based on the expected values of k_1^s and k_1^t, the service capability expectation and the service provider's benefit function are shown in Eqs. (6) and (7), respectively.

$$\begin{cases} E(u_k^{s,i}) = \int_a^b u_k^{s,i} \cdot f(k_1^s) d(k_1^s) \\ E(u_k^{t,j}) = \int_c^d u_k^{t,j} \cdot f(k_1^t) d(k_1^t) \end{cases} \tag{6}$$

$$\begin{cases} E(\pi_k^{s,i}) = \int_a^b \pi_k^{s,i} \cdot f(k_1^s) d(k_1^s) \\ E(\pi_k^{t,j}) = \int_c^d \pi_k^{t,j} \cdot f(k_1^t) d(k_1^t) \end{cases} \tag{7}$$

In order to maximize profits, SSPi and TSPj can maximize the profit by adjusting the service cost coefficient to improve the service quality and ultimately maximize the market share. So, the objective function of SSPi and TSPj is shown in Eq. (8).

$$\begin{cases} \max\{E[\pi_k^{s,i}(u_k^{s,1}, \ldots, u_k^{s,m}, u_k^{t,1}, \ldots, u_k^{t,r-m})]\}, i \in \{1, 2, \ldots, m\} \\ \max\{E[\pi_k^{t,j}(u_k^{s,1}, \ldots, u_k^{s,m}, u_k^{t,1}, \ldots, u_k^{t,r-m})]\}, j \in \{1, 2, \ldots, r-m\} \end{cases} \tag{8}$$

Since the service capabilities of SSPi and TSPj are fixed at a certain moment, this paper maximizes the profit by solving the optimal service capability $(u^{s,i})^*$, $(u^{t,j})^*$ at a certain moment.

$$\begin{cases} (u^{s,i})^* = \arg\{E[\max(\pi_k^{s,i} = f_k^{s,i} \cdot U_k^{s,i})]\}, i \in \{1, 2, \ldots, m\} \\ (u^{t,j})^* = \arg\{E[\max(\pi_k^{t,j} = f_k^{t,j} \cdot U_k^{t,j})]\}, j \in \{1, 2, \ldots, r-m\} \end{cases} \tag{9}$$

Considering that the game parties need to reach the final Nash equilibrium through multiple games, this paper uses Jacobian iterative formula to calculate service capability [9]. Based on the inference process of the reaction function [10], the Eqs. (2), (3) and (4) are transformed to obtain the Jacobian iterative formula of service capability as shown in formula (10).

$$\begin{cases} u_k^{s,i} = \dfrac{P^s - k_2^s q}{2k_1^s} + \dfrac{\sum_{l=1, l \neq i}^{m} u_{k-1}^{s,l} + \sum_{j=1}^{r-m} b_{ij} u_{k-1}^{t,j}}{2a_i} \\ u_k^{t,i} = \dfrac{P^t - k_2^t q}{2k_1^t} + \dfrac{\sum_{h=1, h \neq j}^{r-m} u_{k-1}^{t,h} + \sum_{i=1}^{m} b_{ji}^* u_{k-1}^{s,j}}{2a_j^*} \end{cases} \tag{10}$$

4   Resource Allocation Algorithm

The flowchart of the service capability optimization algorithm is shown in Fig. 2. The algorithm includes: (1) Service provider $SSPi$ and $TSPj$ service capabilities $u_k^{s,i}$ and $u_k^{t,j}$ initialization. (2) For each service provider $SSPi$ and $TSPj$, use the Jacobian iteration formula to solve the optimal service provider service capabilities. (3) Find the expected value of each service provider $SSPi$ and $TSPj$ service capability. (4) Determine the outcome of the competition game until an equilibrium state is obtained. (5) The resource allocation center allocates resources based on the service capabilities of each service provider. The details are described below.

(1) Service provider $SSPi$ and $TSPj$ service capabilities $u_k^{s,i}$ and $u_k^{t,j}$ initialization: When initializing, assign an initial value to each service provider's service capa-bilities. $u_0^{s,i} = v_i^s \in (0, 1)$, $u_0^{t,j} = v_j^t \in (0, 1)$, among them, $i \in \{1, 2, ..., m\}$, $j \in \{1, 2, ..., r - m\}$.

(2) For each service provider $SSPi$ and $TSPj$, use the Jacobian iteration formula to solve the optimal service provider service capabilities: The Eqs. (2), (3), and (4) are transformed to obtain the Jacobian iteration formula (10). The service provider's service capabilities are calculated using Eq. (10) to obtain service capabilities.

(3) Find the expected value of each service provider $SSPi$, $TSPj$ service capability. Use Eq. (6) to find the expected value of service capability $E[u_k^{s,i}]$, $E[u_k^{t,j}]$.

(4) Determine the outcome of the competition game until an equilibrium state is obtained. The expected values of service capabilities $E[u_k^{s,i}]$, $E[u_k^{t,j}]$ are deter-mined. If the convergence condition $\left| E[u_{k+1}^{s,i}] - E[u_k^{s,i}] \right| < \varepsilon \ or \ \left| E[u_k^{s,i}] \right| \geq 1$, $\left| E[u_{k+1}^{t,i}] - E[u_k^{t,i}] \right| < \varepsilon \ or \ \left| E[u_k^{t,i}] \right| \geq 1$ is satisfied, the calculation is stopped and the optimal service capability $(u_k^{s,i})^* = E[u_k^{s,i}]$, $(u_k^{t,j})^* = E[u_k^{t,j}]$ is output, at which time the competition game reaches an equilibrium state. If the condition is not met, the k + 1th iteration is performed until the value of the model Nash equilibrium is obtained.

(5) The resource allocation center allocates resources based on the service capabilities of each service provider. The resource allocation center uses the formula (2) to calculate the $SSPi$, $TSPj$ market ratios $f_k^{s,i}, f_k^{t,j}$ to achieve resource allocation.

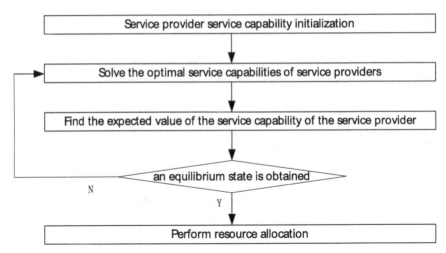

Fig. 2. The flowchart of the service capability optimization algorithm

5 Performance Analysis

In order to verify the results of the algorithm optimization of the power communication network service provider service capabilities. Three competitive game models of competition between *SSP*, competition between *TSP*, and competition between *SSP* and *TSP* were simulated using MATLAB tools. Because the content of the competitive environment is more in line with the actual situation, that is, each service provider cannot know the service cost coefficient of the other party. In the experiment, this environment was compared to an unrestricted environment. Among them, the algorithm in this paper is called the competition game under incomplete information, and the environment without restriction is called the competition game under complete information.

In the simulation environment, the parameter is set to $q = 0.8$, $P^s = 1$, $k_2^s = 1$, $P^t = 0.8$, $k_2^t = 0.8$, $u_0^{s,1} = 0.65$, $u_0^{t,1} = 0.35$. In addition, in the competition game under incomplete information, set $E(k_1^s) = k_1^s = 1$, $E(k_1^t) = k_1^t = 1$, In the competitive game under complete information, $k_1^s = k_1^t = 0.85$. Using the competition between *SSP*1 and *SSP*2 to simulate the competition scenario between *SSP*, set $a_1 = 2.5$, $a_2 = 2.9$, and the experimental results are shown in Fig. 3. It can be seen from the figure that under the incomplete information and the complete information, when the competition starts, *SSP*1 and *SSP*2 do not know the resource input of the other party, so the service capability is not fixed. However, as the number of iterations increases, the service capabilities of *SSP*1 and *SSP*2 gradually converge. The service capabilities of *SSP*1 and *SSP*2 under full information are lower than those under incomplete information. It shows that under incomplete information, both *SSP*1 and *SSP*2 have invested more resources to achieve equilibrium in the competition.

Use the competition between *TSP*1 and *TSP*2 to simulate the competition scene between *TSP*, set $a_1^* = 2.5$, $a_2^* = 2.9$, and the experimental results are shown in Fig. 4. Using the competition between *SSP*1 and *TSP*1 to simulate the competition scenario between *SSP* and *TSP*, set $a_1 = 2.5$, $a_1^* = 2.5$, $b_{11} = 0.5$, $b_{11}^* = 0.6$, and the

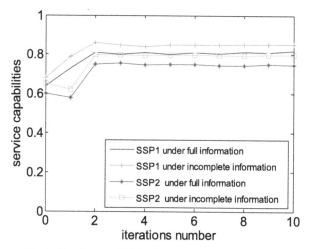

Fig. 3. Service capabilities in a competitive environment

experimental results are shown in Fig. 4 and Fig. 5. As can be seen from the figure, the convergence and resource input of *TSP*1 and *TSP*2, *SSP*1 and *TSP*1 are the same as those in Fig. 3. Therefore, Nash equilibrium is achieved in the two scenarios of competition between *TSP* and competition between *SSP* and *TSP*, and it is more in line with the actual production environment.

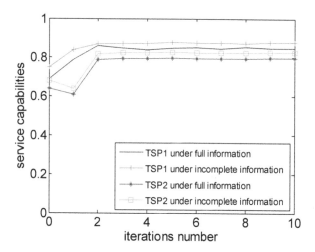

Fig. 4. Service capabilities in a competitive environment

Through the comparative analysis of Fig. 3, 4 and 5, it can be seen that the service optimization algorithm of power communication network service provider in the competitive game environment proposed in this paper can compete between *SSP*, between *TSP*, and between *SSP* and *TSP*. The competition game model is convergent and more realistic than the competition game under full information, so it is more practical for the power company to choose the power communication network service provider.

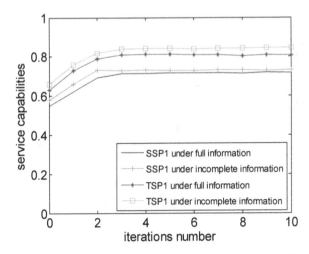

Fig. 5. Service capabilities in a competitive environment between *SSP* and *TSP*

6   Conclusion

With the increasing demand for power communication networks, the network infrastructure built by power companies is gradually unable to meet the development needs of smart grid services. In order to solve this problem, this paper proposes a power communication network resource management architecture, and analyzes how to realize the resource allocation algorithm of power company's revenue maximization under incomplete information. Secondly, the rationality of the model is analyzed through simulation experiments. The simulation results show that the research results of this paper have more practical significance for the power company to choose the power communication network service provider. Although the algorithm is suitable for the real environment, the algorithm only verifies that the model has Nash equilibrium characteristics, and does not analyze the resource utilization in depth. In the next step, based on the model proposed in this paper, we will study how to improve the resource utilization rate of the power communication network.

Acknowledgment. This work was supported by the State Grid Technology Project "Research on Application of interaction between shared mode electric vehicle and power grid"(5418-201971184A-0-0-00) from State Grid Corporation of China.

References

1. Jiang, K., Zeng, Y., Deng, B., et al.: Service-based power communication network risk assessment method. Power Syst. Protect. Control **41**(24), 101–106 (2013)

2. Meng, L., Sun, K., Wei, L., et al.: A virtual resource optimization allocation mechanism for power wireless private network. J. Electron. Inf. Technol. **39**(4), 1–8 (2017)

3. Li, Z., Meng, C.: Resource allocation algorithm of wireless network based on deep reinforcement learning. Commun. Technol. **53**(08), 1913–1917 (2020)

4. Li, M., Xu, Z., Xu, C., et al.: QoS-driven power communication network utility maximizes resource allocation mechanism. Comput. Syst. Appl. **27**(7), 265–271 (2018)

5. Hu, Y., Zhuang, L., Chen, H., et al.: Bayesian virtual network mapping of time and energy perception. J. Commun. **37**(6), 106–118 (2016)

6. Meng, C., Wang, G., Dai, X., et al.: An energy-efficient transmission strategy for cache-enabled wireless networks with non-negligible circuit power. IEEE Access **7**, 74811–74821 (2019)

7. Zhang, H., Zhang, Z., Long, K.: NOMA heterogeneous network resource allocation based on mobile edge computing. J. Commun. **41**(4), 27–33 (2020)

8. Liu, L., Zhang, D.: Network resource demand forecasting algorithm for mission service quality. Ship Electron. Eng. **39**(11), 127–130 (2019)

9. Tang, C., Dong, S., Ren, X., et al.: Improved Jacobi preprocessing method for iterative power flow calculation. Autom. Electr. Syst. **42**(12), 81–86 (2018)

10. Dai, Y., Gao, H., Gao, Y., et al.: Smart grid real-time pricing mechanism with power demand forecast update. Autom. Electr. Syst. **42**(12), 58–63 (2018)

Security and Stable Control

Multi-domain Cooperative Service Fault Diagnosis Algorithm Under Network Slicing with Software Defined Networks

Wei Li[1], Yong Dai[1], Yong Xu[1], Xilao Wu[1], Wei Li[1], and Peng Lin[2(✉)]

[1] Communication Branch, State Grid Jiangsu Electric Power Co., Ltd., Jiangsu, China
[2] Beijing Vectinfo Technologies Co., Ltd., Beijing, China
linpeng@vectinfo.com

Abstract. In order to solve the problem of low accuracy of fault diagnosis algorithms in multiple management domain environments such as such as Software Defined Networks (SDN), this paper proposes a multi-domain cooperative service fault diagnosis algorithm under network slice based on the correlation between faults and symptoms. According to the relationship between the management domain and the symptoms, the network resources corresponding to the symptoms are divided into resources within the management domain and inter-domain resources. When constructing a suspected fault set, the suspected fault set is constructed according to the number of simultaneous faults, and the final suspected fault set is determined by calculating the interpretation capability of the suspected fault. Finally, according to Bayesian theory, the fault set with the highest probability is regarded as the most probable fault set. Compared with the existing classical algorithms in the experimental part, it is verified that the algorithm in this paper improves the accuracy of fault diagnosis and reduces the false alarm rate of fault diagnosis.

Keywords: SDN network · Network slicing · Fault diagnosis · Management domain

1 Introduction

With the rapid construction and operation of next generation networks, the application scope of various network-based services in production and life is gradually increasing. In order to improve the reliability of the network, network virtualization technology such as Software Defined Networks (SDN) has been applied to 5G networks [1]. In this context, existing networks are divided into underlying networks and virtual networks. The underlying network is responsible for the construction of the underlying network nodes and the underlying network links. The virtual network leases network resources from the underlying network to run specific 5G services. When network resources fail, how to quickly and accurately locate the fault has become a key issue that network operators urgently need to solve.

© ICST Institute for Computer Sciences, Social Informatics and Telecommunications Engineering 2021
Published by Springer Nature Switzerland AG 2021. All Rights Reserved
M. Cheng et al. (Eds.): SmartGIFT 2020, LNICST 373, pp. 55–64, 2021.
https://doi.org/10.1007/978-3-030-73562-3_5

The network fault diagnosis algorithm mainly adopts two strategies: passive detection [2] and active detection [3]. The main advantage of passive detection is simple implementation, and the main disadvantage is the low accuracy of the fault diagnosis model constructed. Active detection can better improve the performance of the fault diagnosis algorithm by selecting the detection strategy in advance, but the design of detection is more complicated. For example, literature [4] uses a dependency matrix to construct a detection model, which better solves the problem of single-point fault diagnosis. In terms of multi-layer fault diagnosis, the general method is to resolve the multi-layer model into a two-layer model based on the network resource relationship [5]. For the problems of complex network topology and low performance of fault diagnosis algorithms brought about by the large-scale network, literature [6] uses artificial intelligence algorithms to construct learning models, which better solve the problem of low performance of fault diagnosis algorithms in large-scale environments.

The existing research mainly solves the fault location in a single domain. However, when the network scale becomes larger and larger, multiple network operators will jointly build and manage the network, thereby forming multiple management domains. Each domain is responsible for network resource allocation and fault management in the area. When a virtual network service fails, each domain only knows its own internal failure information. When faults cannot be located within a domain, the problem of how multiple domains can collaborate to locate faults has not been well resolved. To solve this problem, this paper proposes a multi-domain cooperative service fault diagnosis algorithm under network slicing with SDN. The algorithm improves the performance of the fault diagnosis algorithm through the cooperation of multiple management domains.

2 Problem Description

Network slicing is an on-demand networking method that allows operators to separate multiple virtual end-to-end networks on a unified infrastructure. Each network slicing is carried out from the wireless access network, the bearer network to the core network. In a network slice, it can be divided into at least three parts: wireless network sub-slice, bearer network sub-slice and core network sub-slice. The core of network slicing technology is network function virtualization. Network function virtualization separates the hardware and software parts from traditional networks. The hardware is deployed by a unified server, and the software is undertaken by different network functions, thereby realizing the needs of flexible assembly services. Network slicing is based on a logical concept and is the reorganization of resources. Reorganization is to select the required virtual machines and physical resources for a specific communication service type according to the service level agreement.

In the network slicing environment, in order to distinguish the existing network from the sliced network resources, the physical network resource is called the underlying network, and each sliced resource is called the virtual network. Use undirected weighted graph $G^S = (N^S, E^S)$ to represent the underlying network. Use undirected weighted graph $G^V = (N^V, E^V)$ to represent the virtual network. $n_i^s \in N^S$ and $n_i^V \in N^V$ represent the underlying node and virtual node, respectively, and $e_j^s \in E^S$ and $e_j^V \in E^V$ represent the underlying link and the virtual link, respectively. Because the virtual network allocates resources by the underlying network, use $Mapping_N : (N^V \to N^S, E^V \to P^S)$

to represent the resource allocation relationship between the virtual network and the underlying network. Among them, $N^V \rightarrow N^S$ indicates that the underlying node n_i^s allocates resources to the virtual node n_i^V, and $E^V \rightarrow P^S$ indicates that the underlying path P^S allocates resources to the virtual link e_j^V. The bottom-level path P^S refers to the bottom-level link resource composed of multiple end-to-end connected bottom-level links e_j^s. The start and end points of the path respectively correspond to the bottom-level nodes mapped by the two virtual nodes of the virtual link.

When the virtual network covers a large area, multiple domains need to cooperate with each other to meet the resource requirements of the virtual network. The service model of multi-domain collaboration is shown in Fig. 1. It contains 3 management domains. The virtual network uses the network resources of these three management domains to construct a virtual network. When a virtual resource on a virtual network fails, the cooperation of the three management domains is required to quickly locate the root cause of the failure.

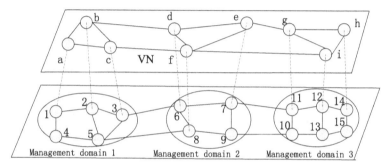

Fig. 1. Multi-domain collaboration service model

3 Fault Propagation Model

Because all users want a reliable network, fault management is one of the most basic functions of network management. When a component in the network fails, the network manager must quickly find the root cause of the fault and troubleshoot it in time. Under normal circumstances, it is unlikely that a fault can be quickly isolated, because the factors that cause network faults are usually very complex, especially those caused by multiple network components. In this situation, we should generally repair the network first, and then analyze the cause of the failure. The recurrence of similar failures can be prevented by analyzing the causes of failures, which is very important for the reliable performance of the network. The goal of fault management is to resume normal service operations as soon as possible, minimize the negative impact of component failure on the business, and ensure that the service level goals and service level quality agreed with the business customers in advance are met.

In order to quickly locate faults, a fault propagation model is constructed based on Bayesian theory, so as to correlate the observed symptoms with the actual network environment. The Bayesian network is a directed acyclic graph G(V, E). The node V in the graph represents a variable, and the directed edge E connecting the nodes represents a dependency between the nodes. Each node stores a conditional probability table, which indicates the influence of the value of its parent node on the state of the node. If the node is the root node, the conditional probability of the node records the prior probability of this node. The fault propagation model is shown in Fig. 2, including symptom, fault, and directed line from fault to symptom.

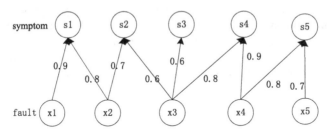

Fig. 2. Fault propagation model based on Bayesian theory

Symptoms refer to the working status of various businesses running on the virtual network. Symptom set $S_o = \{s_1, s_2, ..., s_m\}$ represents a set of m symptoms. When the business is running normally, it is called a positive symptom and is represented by $s_m = 0$. When the business fails to operate normally, it is called a negative symptom and is represented by $s_m = 1$. Failure refers to the working status of the underlying network resources. The set of suspected faults $X = \{x_1, x_2, ..., x_n\}$ represents a set of n suspected faults x. When the underlying network resources are operating normally, use $x_n = 0$ to indicate. When the underlying network resources are abnormal, use $x_n = 1$ to indicate. The directed line from failure to symptom indicates the probability that when the underlying network resource is abnormal, the symptom status of the service carried on the underlying network resource is negative.

4   Algorithm

This paper proposes a multi-domain cooperative service fault diagnosis algorithm under network slice (MCSFDA) with SDN as shown in Fig. 1. The algorithm includes the following three processes. (1) Symptom collection and fault decomposition, (2) Building Bayesian model, (3) Fault set location. In step (1), each virtual network service provider reports symptoms and fault information to the fault management center, and the fault center performs fault decomposition based on the collected symptoms and network topology, and sends the fault information to the corresponding management domain. In step (2), each management domain uses detection technology to obtain the network performance of each faulty node, and feeds the results back to the fault management center. For links between domains, related domains need to send data packets to each other to obtain the packet loss rate of the link. For example: in Fig. 1, the path (1–2–3) belongs to domain 1, the link (6–7) belongs to domain 2, the path (11–12–14) belongs to domain 3, and the link (3–6) belongs to the shared resources of domain 1 and domain 2, links (7–11) belong to the shared resources of domain 2 and domain 3. The fault management center builds a Bayesian model based on the network performance fed back from each management domain. In the fault propagation model based on Bayesian theory, the symptom refers to the status of the virtual network service, and the fault refers to the detection result of the underlying link corresponding to the virtual network service. In step (3), to select the most suitable fault set from the set of suspected faults to realize fault location. It adopts two processes: constructing a set of suspected faults and locating faults based on Bayesian formula. Steps (1) and (3) are described in detail below.

4.1 Symptom Collection and Fault Decomposition

The end-to-end service $P^V\left(n_{p1}^V, n_{pm}^V\right)$ contains multiple virtual paths. From the process of mapping $E^V \to P^S$ from the virtual link to the underlying path, it can be seen that the end-to-end service $P^V\left(n_{p1}^V, n_{pm}^V\right)$ contains more underlying links. To facilitate the description of the underlying links included in the end-to-end service, it is necessary to map the end-to-end service to the underlying link. Use $e^V\left(n_k^V, n_l^V\right)$ to represent the virtual link between virtual nodes n_k^V and n_l^V, and use $P^V\left(n_{p1}^V, n_{pm}^V\right)$ to represent the virtual path between virtual nodes n_{p1}^V and n_{pm}^V. $P^V\left(n_{p1}^V, n_{pm}^V\right)$ uses link to represent $e^V\left(n_{p1}^V, n_{p2}^V\right)$, $e^V\left(n_{p2}^V, n_{p3}^V\right), \ldots, e^V\left(n_{pm-1}^V, n_{pm}^V\right)$. Use $e^S\left(n_k^S, n_l^S\right)$ to represent the underlying link between the underlying nodes n_k^S and n_l^S, and use $P^S\left(n_{p1}^S, n_{pm}^S\right)$ to represent the underlying path between the underlying nodes n_{p1}^S and n_{pm}^S. $P^S\left(n_{p1}^S, n_{pm}^S\right)$ use link to represent $e^S\left(n_{p1}^S, n_{p2}^S\right)$, $e^S\left(n_{p2}^S, n_{p3}^S\right), \ldots, e^S\left(n_{pm-1}^S, n_{pm}^S\right)$.

According to the relationship of $E^V \to P^S$, convert $P^V\left(n_{p1}^V, n_{pm}^V\right)$ to $P^S\left(n_{p1}^S, n_{pm}^S\right)$. If the fault in $P^S\left(n_{p1}^S, n_{pm}^S\right)$ can be inferred based on the symptoms, the faulty resources can be repaired to ensure the quality of service. However, when the underlying link contained in $P^S\left(n_{p1}^S, n_{pm}^S\right)$ is provided by multiple underlying network resource management domains, multiple management domains need to cooperate with each other to complete fault diagnosis. Taking into account that each management domain can detect the failure of its own internal network resources, this paper divides $P^S\left(n_{p1}^S, n_{pm}^S\right)$ into the resources of path $P^S\left(I_{i,j}^{i+1}, E_{i,j}^{i+2}\right)$ in the management domain and inter-domain link $e\left(E_{i,j}^i, I_{i,j}^{i+1}\right)$ according to the characteristics of the management domain. Among them, $I_{i,j}^k$ represents the ingress gateway of the k-th SN_k, and $E_{i,j}^k$ represents the egress gateway of the k-th SN_k. For $I_{i,j}^k$, the constraints of $n_{p1}^S \in SN_i$, $n_{pm}^S \in SN_j$, $n_{pn}^S \in SN_k$ and $n_{pn-1}^S \notin SN_k$ should be satisfied. For $E_{i,j}^k$, the constraints of $n_{p1}^S \in SN_i$, $n_{pm}^S \in SN_j$, $n_{pn}^S \in SN_k$ and $n_{pn+1}^S \notin SN_k$ should be satisfied. Therefore, $P^S\left(n_{p1}^S, n_{pm}^S\right)$ can be expressed as

$e^S\left(n_{p_1}^S, E_{i,j}^i\right), e\left(E_{i,j}^i, I_{i,j}^{i+1}\right), p^S\left(I_{i,j}^{i+1}, E_{i,j}^{i+2}\right), \ldots, e\left(E_{i,j}^{j-1}, I_{i,j}^j\right)$. For example, the underlying network of the end-to-end service (a-b-d-e-g-h) in Fig. 1 can be divided into: the path in domain 1 (1–2–3), the inter-domain link (3–6), the path in domain 2 (6–7), the inter-domain link (7–11), and the path in domain 3 (11–12–14).

4.2 Fault Location

Fault set location includes two processes: constructing a set of suspected faults and locating faults based on Bayesian formula. When constructing a set of suspected failures, construct a set of suspected failures based on the number of simultaneous failures. The number of simultaneous failures is related to the number of network nodes and the probability of network node failures. Assuming that the probability of failure of the underlying network is 0.001 and network is composed of three network nodes, the probability of failure of two nodes at the same time is 3×10^{-6}. When constructing a set of suspected failures, a failure node is arbitrarily selected from the failure node set X, and placed into the candidate failure set m_{ik} (i represents the size of the candidate failure set, and k represents the sequence number of the set). The failure set $M_i(i = 1, \ldots, \max(\Omega, |X|))$ is gradually constructed until the end condition is met, that is, Ω nodes are included in m_{ik}. Among them, Ω represents the number of simultaneous failures.

In order to evaluate the value of m_{ik}, define $Abli_{ik}$ as the explanatory ability of m_{ik}. The calculation method is shown in formula (1). $Num_{A(s)=1}$ indicates the number of abnormal symptoms related to the suspected failure; $A(s) = 1$ indicates that the symptom s is abnormal, and the calculation formula is shown in formula (2). $Num_{A(s)=0}$ indicates the number of normal symptoms related to the suspected failure; $A(s) = 0$ indicates that the symptom s is abnormal, and the calculation formula is shown in formula (3). $pa(s)$ represents the parent node of symptom s.

$$Abli_{m_{ik}} = Num_{A(s)=1} + Num_{A(s)=0} \tag{1}$$

$$Num_{A(s)=1} = |\{s | s \in S_o, A(s) = 1, \exists x \in m_{ik}, x \in pa(s)\}| \tag{2}$$

$$Num_{A(s)=0} = |\{s | s \in S_o, A(s) = 0, \exists x \in m_{ik}, x \in pa(s)\}| \tag{3}$$

According to Bayesian theory, if the state of some nodes is known, formula (4) can be used to solve the maximum possible state of unknown node $X = \{X_1, X_2, \ldots, X_n\}$. Among them, $pa(T_j)$ represents the parent node of detecting T_j. N represents the number of nodes, and M represents the number of probes.

$$\begin{aligned} X^* &= \max_X P(X|T) = \max_X \frac{P(X,T)}{P(T)} = \max_X P(X,T) \\ &= \max_X \prod_{i=1}^N P(X_i) \prod_{j=1}^M P(T_j|pa(T_j)) \end{aligned} \tag{4}$$

Therefore, this paper uses formula (5) to calculate the probability of each $M_i(i = 1, \ldots, \max(\delta, |Fs|))$, and regards the failure set with the largest probability as the most likely failure set. The nodes included in the failure set are the failed nodes.

$$P(M_i) = \prod_{X_i \in X} P(X_i) \prod_{T_j \in T} P(T_j|pa(T_j)) \tag{5}$$

5 Performance Analysis

5.1 Network Environment

In order to simulate the network topology in the network slicing environment, this article uses the GT-ITM [7] tool to generate the underlying network and virtual network topology to simulate the network slicing environment. In order to judge the performance of the algorithm in different network environments, the node size of the underlying network was increased from 100 to 500. The number of virtual nodes in the virtual network is uniformly distributed from 5 to 25, which is used to simulate virtual networks of different sizes. The mapping algorithm from the underlying network to the virtual network uses the classic mapping algorithm [8]. In order to simulate different management domains, the bottom network will be divided into 5 management domains according to the number of bottom network nodes. In terms of service simulation, this article takes end-to-end service as the research object. Select 10% of the virtual nodes from the virtual network as the source node. For each virtual source node, 3 nodes are randomly selected as destination nodes, and the shortest path algorithm is used to generate end-to-end services. In terms of fault injection, set the prior failure probability of the underlying network node to obey the uniform distribution of [0.001,0.01], and the conditional probability obey the uniform distribution of (0,1).

In order to analyze the performance of the algorithm MCSFDA in this paper, it is compared with the non-cooperative service fault diagnosis algorithm (NCSFDA). Different from the algorithm in this paper, each management domain of the algorithm NCSFDA sends the network performance to the management center, and the management center directly diagnoses the fault based on the mapping relationship between the virtual network and the underlying network. The evaluation indicators include the accuracy rate of fault diagnosis, false alarm rate, and diagnosis time. The accuracy rate refers to the proportion of the diagnosed faulty node set in the real faulty node set. The higher the accuracy rate, it means that the algorithm has identified more real faults and the algorithm performance is better. The false alarm rate refers to the proportion of false faults identified by the diagnostic algorithm in all identified faults. The higher the false alarm rate, it indicates that the algorithm mistakenly recognizes the normal network node as the fault node, and the performance is poor. Diagnosis time refers to the time taken by the algorithm from inputting network topology and service symptom information to outputting diagnosis results. The longer the diagnosis algorithm takes, the greater the time overhead of the algorithm.

5.2 Performance Comparison

The accuracy of fault diagnosis is shown in Fig. 3. The X-axis indicates that the number of network nodes has increased from 100 to 500, which is used to analyze the impact of different network sizes on algorithm performance. The Y axis represents the accuracy of the algorithm. It can be seen from the figure that the size of the network has a small effect on the accuracy of the fault diagnosis of the two algorithms, indicating that the diagnosis performance of the two algorithms has little relationship with the network topology. From the comparison of the accuracy of the two algorithms, the accuracy of this algorithm is high. This is because the algorithm in this paper can effectively improve the accuracy of the data in the fault diagnosis model through the collaboration of multiple domain managers, thereby improving the accuracy of fault diagnosis.

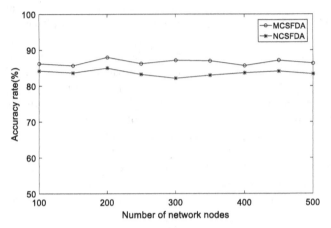

Fig. 3. Comparison of accuracy rate

The comparison result of the false alarm rate of fault diagnosis is shown in Fig. 4. The X axis represents the number of network nodes, and the Y axis represents the false alarm rate of the algorithm. It can be seen from the figure that the false alarm rate performance of the two algorithms has little to do with the network scale. The false alarm rate of the algorithm in this paper is lower than that of the traditional algorithm. The same is because the fault diagnosis model data of the algorithm in this paper is more accurate, which reduces the false alarm rate.

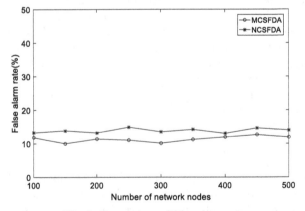

Fig. 4. Comparison of false alarm rate

The comparison of the duration of fault diagnosis is shown in Fig. 5. The X axis represents the number of network nodes, and the Y axis represents the fault diagnosis time of the algorithm. It can be seen from the figure that as the network scale increases, the diagnosis time of the two algorithms is increasing. This is because as the network scale increases, the fault propagation model increases, and the set of suspected faults also increases, which requires more time for fault diagnosis. In addition, the diagnosis time of the algorithm in this paper has increased rapidly. This is because, compared with traditional algorithms, it requires cooperation between various domains for active positioning, which requires a longer time overhead.

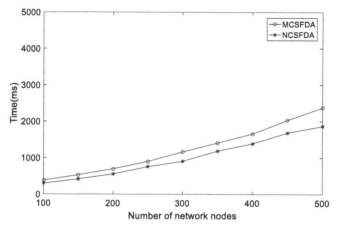

Fig. 5. Comparison of time

6   Conclusion

Accurate and rapid location of SDN network resource failures has become a key issue that network operators urgently need to solve. However, when network resources are composed of multiple management domains, the accuracy of service fault diagnosis across multiple domains is low. To solve this problem, this paper proposes a multi-domain cooperative service fault diagnosis algorithm under 5G network slicing with SDN. The algorithm includes three processes: symptom collection and fault decomposition, Bayesian model construction, and fault set location. In the symptom collection and fault decomposition steps, the fault center performs fault decomposition based on the collected symptoms and network topology, and sends the fault information to the corresponding management domain. In the step of constructing the Bayesian model, inter-domain links need related domains to send data packets to each other to obtain the packet loss rate of the link. In the fault set locating step, it includes two processes: constructing a set of suspected faults and locating faults based on Bayesian formula. The algorithm uses detection technology to obtain network performance through the collaboration of various management domains, and builds a Bayesian model and fault location based on network performance. Compared with existing research, this paper improves the performance of fault diagnosis through fault decomposition and management domain collaboration.

With the increase of network scale and the development of artificial intelligence technology, how to further realize autonomous large-scale fault diagnosis is an important and feasible research topic. In the next step, based on the research results of this article, we will explore a large-scale network fault autonomous diagnosis system based on artificial intelligence. Further improve the availability and convenience of the fault diagnosis system.

Acknowledgement. This work is supported by science and technology project from State Grid Jiangsu Electric Power Co., Ltd: "Technology Research for High-efficiency and Intelligent Cooperative Wide-area Power Data Communication Networks (SGJSXT00DDJS1900168)".

References

1. Lun, T., Yu, Z., Qi, T., et al.: 5G network slicing virtual network function migration algorithm based on reinforcement learning. J. Electron. Inf. Technol. **42**(3), 669–677 (2020)
2. Dusia, A., Sethi, A.S.: Recent advances in fault localization in computer networks. IEEE Commun. Surv. Tutor. **18**(4), 3030–3051 (2016)
3. Wu, B., Ho, P.H., Tapolcai, J., et al.: Optimal allocation of monitoring trails for fast SRLG failure localization in all-optical networks. In: Proceedings of 2010 IEEE Global Telecommunications Conference, Miami, USA, pp. 1–5 (2010)
4. Brodie, M., Rish, I., Ma, S., et al.: Active probing strategies for problem diagnosis in distributed systems. In: Proceedings of the 18th International Joint Conference on Artificial Intelligence, pp. 1337–1338. Acapulco, Mexico (2003)
5. Ogino, N., Kitahara, T., Arakawa, S., et al.: Decentralized Boolean network tomography based on network partitioning. In: Proceedings of 2016 IEEE/IFIP Network Operations and Management Symposium, Istanbul, Turkey, pp. 162–170 (2016)
6. Srinivasan, S.M., Tram, T.-H., et al.: Machine learning-based link fault identification and localization in complex networks. IEEE Internet Things J. **6**(4), 6556–6566 (2019)
7. Zegura, E.W., Calvert, K.L., Bhattacharjee, S.: How to model an internetwork. In: Proceedings of IEEE INFOCOM (1996)
8. Yu, M., Yi, Y., Rexford, J., Chiang, M.: Rethinking virtual network embedding: substrate support for path splitting and migration. ACM SIGCOMM CCR. **38**(2), 17–29 (2008)

Fault Diagnosis Algorithm Based on Service Characteristics Under Software Defined Network Slicing

Wei Li[1], Hao Cai[1], Chunxia Jiang[1], Ping Xia[1], Song Jiang[1], and Peng Lin[2(✉)]

[1] Communication Branch, State Grid Jiangsu Electric Power Co., Ltd., Jiangsu, China
[2] Beijing Vectinfo Technologies Co., Ltd., Beijing, China
linpeng@vectinfo.com

Abstract. In order to solve the problem of low accuracy of fault diagnosis algorithms brought by network dynamics, this paper proposes a fault diagnosis algorithm based on service characteristics under software defined network slicing. In order to reduce the problem of inaccurate symptom information caused by network dynamics, the credibility of symptoms is calculated based on the alternative probabilistic characteristics of network nodes, and the symptom information is corrected. The node importance is analyzed from the two dimensions of node centrality and number of links. Based on the node importance and symptom information, the reliability of the node failure is ranked. Finally, based on the maximum coverage algorithm, the optimal set of suspected faults is selected from the set of suspected faults as the final set of faults. The experiment compares the algorithm in this paper with the existing algorithm, and verifies that the algorithm in this paper effectively improves the accuracy of fault diagnosis.

Keywords: Software defined network · Network slicing · Fault diagnosis · Network characteristics

1 Introduction

With the rapid construction and operation of future networks, the advantages of 5G networks such as high bandwidth, low latency, and high number of connections have become more prominent, and the types and number of services have increased rapidly. Taking into account the different requirements of different services on the network and the rapid increase of new service types, network slicing technology with Software Defined Networks (SDN) has been proposed and gradually accepted by global network operators and equipment vendors and has become the mainstream technology [1]. In the network slicing environment, the traditional physical network is divided into the underlying network and the virtual network. The underlying network provider is responsible for building the underlying nodes and underlying links. The virtual network service provider leases the underlying nodes and links from the underlying network to construct a virtual network.

© ICST Institute for Computer Sciences, Social Informatics and Telecommunications Engineering 2021
Published by Springer Nature Switzerland AG 2021. All Rights Reserved
M. Cheng et al. (Eds.): SmartGIFT 2020, LNICST 373, pp. 65–76, 2021.
https://doi.org/10.1007/978-3-030-73562-3_6

Various services are carried on the virtual network. In order to ensure the stable and reliable operation of 5G services, fast and accurate fault location technology has become an important task for network managers.

From the proactive perspective of fault diagnosis, fault diagnosis algorithms can be divided into two types: active diagnosis and passive diagnosis. Active diagnosis is fault location of specific network resources based on the network environment and business characteristics, through the selection of detection sites and implementation of detection [2]. Passive diagnosis means that after network managers receive network alarms, they locate faults based on the alarm information and network information [3]. From the perspective of the mathematical model of fault diagnosis, fault diagnosis algorithms can be divided into binary dependent fault models, graph dependent fault models, Bayesian dependent matrix models, etc. [4]. By constructing a fault model and using mathematical tools, the complexity of fault diagnosis can be better solved [5]. Aiming at the fault diagnosis problem in the network virtualization environment, literature [6] proposed a rule-based hybrid tracking mechanism, which better solved the problem of low fault diagnosis performance caused by heterogeneous networks in the SDN environment.

Based on the correlation between faults and symptoms, existing studies have achieved good results in the accuracy rate of fault diagnosis, false alarm rate, diagnosis time and other indicators by constructing fault diagnosis models. However, because software defined network slicing technology realizes the dynamic migration and on-demand allocation of resources, the relationship between faults and symptoms is more complicated, resulting in a decrease in the accuracy of the fault diagnosis model, thereby affecting the performance of fault diagnosis. In order to solve the problem of low performance of fault diagnosis algorithms caused by network dynamics, this paper proposes a fault diagnosis algorithm based on service characteristics under software defined network slicing. Based on the relationship between network resources and services, the algorithm corrects symptom information and ranks the reliability of node failures. In the experimental link, it is verified that the algorithm in this paper effectively improves the performance of the fault diagnosis algorithm.

2   Problem Description

The biggest difference between network slicing and the current Internet is that the current Internet is composed of a single role of an Internet service provider, while network slicing is composed of two different roles, an infrastructure provider and a network service operator. This division of labor realizes the separation of network equipment and terminal services, and facilitates the deployment of users' target requirements on network slicing. In addition, each network slicing can use independent protocol architectures, and can dynamically adjust and re-allocate node resources and link resources in the network according to user needs, so as to achieve efficient and reasonable use of network resources and improve network performance. Service quality reduces network operation and maintenance costs. The business model of network slicing mainly includes three entities: infrastructure providers, network service operators, and end users.

In the network slicing environment, the infrastructure provider is responsible for the construction and maintenance of the underlying physical network, ensuring the normal operation of the physical network, and providing physical resources to different network service operators through open and standard programmable interfaces. Different infrastructure providers provide differentiated services in terms of resource quality and freedom of use. Multiple infrastructure providers communicate and combine based on mutually agreed interconnection protocols to create an end-to-end physical architecture. Among them, the task of some infrastructure providers is to use different network technologies to provide network service operators with connection services, such as optical fiber, satellite, etc., which are called equipment providers. Another part of infrastructure provider is responsible for connecting customer premises equipment to the core network and is called access provider.

By purchasing or renting physical network resources provided by infrastructure providers, network service operators build virtual networks according to their own needs and provide them to end users, and network service operators can also rent their own resources to other network service operators. Therefore, the resources of the virtual network may come from multiple types of providers. Network service operators are mainly responsible for the creation, modification and cancellation of virtual networks, and have the ability to monitor resources on the virtual network, obtain network operating parameters in real time, and detect failures.

End users in the network virtualization environment are similar to current users on the Internet, but they have more choices. Each terminal user can establish different connections with one or more network service operators at the same time through the mapping agent, use the network protocol of the virtual network to access the services provided by the virtual network, and obtain matching services according to business requirements.

In the software defined network slicing environment, use $G_S = (N_S, E_S)$ to represent the underlying network topology and $G_V = (N_V, E_V)$ to represent the virtual network topology. N_S and E_S respectively represent the bottom layer node set and the bottom layer link set. N_V and E_V respectively represent a set of virtual nodes and a set of virtual links. Suppose a service $s_i \in S$ is running on a virtual network. If service $s_i \in S$ is abnormal, the cause of the failure generally includes service software failure and underlying network failure. If it is a service software failure, service providers on the virtual network can quickly resolve service abnormalities through strategies such as software reconfiguration or software upgrades.

However, if the cause of the failure is an underlying network failure, the virtual network operator needs to report the service abnormality to the underlying network operator, and the underlying network operator will locate the fault. Fault management in the network management system is the core function of network management. Fault management is a series of management activities for abnormal situations in the managed network and its environment. Its work includes timely and accurate detection and location of faults, activation of fault control functions to eliminate and isolate faults or perform fault recovery. Fault management includes three stages: fault detection, fault diagnosis and fault recovery. The purpose of fault detection is to monitor whether a fault occurs in the network by collecting and analyzing network data. The purpose of fault diagnosis is to use further means to determine the root cause of the fault when a fault occurs in the network. Fault recovery is to perform corresponding fault correction measures on the faulty node after finding the root cause of the fault.

This article mainly studies how the underlying network operator can quickly locate faults based on the service exception information reported by the virtual network operator. Considering that the underlying network operator has real-time virtual network resource allocation information to the underlying network, how to map the virtual network resources used by the service to the underlying network resources is a relatively easy problem for the underlying network service provider. This article mainly studies the fault diagnosis between the service and the underlying network.

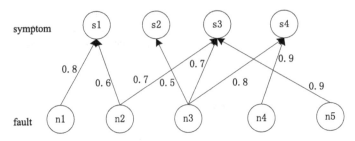

Fig. 1. Fault propagation model based on Bayesian network

Based on the fault data of the service and the underlying network, the fault diagnosis model is constructed as shown in Fig. 1. It can be seen from the figure that the Bayesian network-based fault propagation model includes symptom nodes, fault nodes, and directed lines from fault nodes to symptom nodes. A faulty node refers to the probability of a network node failure, denoted by $p(n_j)$. The symptom node refers to the probability that the service is abnormal, denoted by $p(s_j)$. The directed line from the fault node to the symptom node refers to the probability that the symptom of the service s_i is abnormal when the underlying node n_j fails, and is represented by $p(s_i|n_j)$.

This article mainly solves the problem of how to quickly infer the location of the underlying network failure after the underlying network operator receives the service exception information. Using $p(n_j|s_i)$ to represent the probability of failure of network node n_j based on an extrapolation of the known state of service s_i. Based on Bayesian inference, $p(n_j|s_i)$ can be calculated using formula (1). From the formula (1), we can see that to calculate the root cause failure of all abnormal services, we need to know the values of $p(n_j)$, $p(s_i)$, and $p(s_i|n_j)$. The more accurate these values, the more accurate the inferred underlying network node.

$$p(n_j|s_i) = \frac{p(s_i|n_j) \cdot p(n_j)}{\sum_{n_i \in N_s} p(s_i|n_j) \cdot p(n_j)} \tag{1}$$

Generally speaking, the values of $p(n_j)$, $p(s_i)$, and $p(s_i|n_j)$ can be obtained based on long-term network operation data. However, considering the software defined network environment, the network becomes more and more dynamic. Therefore, how to optimize the values of $p(n_j)$, $p(s_i)$, and $p(s_i|n_j)$ has become an effective measure to improve the performance of the fault diagnosis algorithm. For ease of description, the following describes the values of $p(n_j)$, $p(s_i)$, and $p(s_i|n_j)$ in a formal way. For a faulty node, if $p(n_j) = 1$ it means that node n_j has a fault, it is called a faulty node; if $p(n_j) = 0$, it means that node n_j has no fault, it is called a faultless node; when $0 \le p(n_j) \le 1$, it is called a suspected fault node. For the symptom node, if $p(s_i) = 1$, it means that the status of service s_i is abnormal, which is called negative symptom; if $p(s_i) = 0$, it means that the state of service s_i is normal, which is called positive symptom; when $0 \le p(s_i) \le 1$, it is called a suspected negative symptom node.

3 Fault Propagation Model

It can be seen from the fault propagation model that faults and symptoms are the main elements of the fault propagation model. If you want to improve the performance of the fault diagnosis algorithm, the best strategy is to ensure the accuracy of the fault and symptom data. Based on this, this section analyzes and optimizes the key elements of the fault propagation model from the two dimensions of symptom optimization and fault sequencing.

3.1 Symptom Optimization

Because of its high reliability and saving network resources, dynamic routing has gradually become the main routing protocol. However, the dynamic routing strategy can easily cause the faulty node to be replaced by the available node, and the network management system cannot detect the real fault. For negative symptoms, there must be a faulty node in the node that it passes through. For a positive symptom, there is a situation where an available node is used instead of a faulty node, so there may be a faulty node in the symptom. Based on this, the steps of the optimization method for the probability of positive symptoms adopted in this paper include three sub-processes: calculating the alternative probability of each node, sorting according to the alternative probability, and symptom optimization.

According to the operation process of the dynamic routing protocol, if a node fails, the dynamic routing protocol will use the available nodes near the failed node to replace it. In order to evaluate whether a node has a substitute node, this paper uses the two-hop distance correlation of the node for evaluation. Assuming that there are nodes n_j and n_k, formula (2) can be used to calculate the two-hop distance correlation between these two nodes. $Q(n_j)$ represents the set of nodes connected to the network node n_j. It can be seen from formula (2) that the more similar the adjacent network topology of nodes n_j and n_k, the higher the possibility that nodes n_j and n_k will replace each other.

$$J(n_j, n_k) = \begin{cases} \dfrac{|Q(n_j) \cap Q(n_k)|}{|Q(n_j) \cup Q(n_k)|} & n_j \text{ and } n_k \text{ are not directly connected} \\ 1 & n_j \text{ and } n_k \text{ are directly connected} \end{cases} \quad (2)$$

Based on the probability of mutual substitution between network nodes, formula (3) is used to calculate the alternative probability of a node. Formula (3) is used to solve the alternative probability of network node n_j in the network topology, where n_j^a and n_j^b represents the neighboring point of network node n_j.

$$S(n_j) = \sum_{n_j^a, n_j^b \in Q(n_j)} \left(1 - J\left(n_j^a, n_j^b\right)\right) \tag{3}$$

From the analysis of dynamic routing protocol characteristics and the calculation process of node substitution probability, it can be seen that the higher the substitutability of the node, the lower the credibility of the positive symptoms. The lower the substitutability, the higher the credibility of the positive symptoms. Therefore, according to the ascending order of the alternative probability of the network nodes, the network node set N_S^{rep} is obtained. The credibility of the positive symptoms related to the nodes ranked first in the network node set N_S^{rep} is higher. In the fault diagnosis model, the optimization method for the positive symptom connected to node n_j is formula (4). Among them, $S_{n_j}^o$ represents the o-th positive symptom node connected to the network node n_j.

$$\xi_{n_j}^o = S(n_j)^* p\left(s_{n_j}^o\right) \tag{4}$$

For the convenience of analysis, $\xi_{n_j}^0$ is normalized using formula (5), and the result is $\xi_{n_j}^{0-nor}$, where ξ represents the set formed by $\xi_{n_j}^0$, $\max_\xi$ represents the maximum value in the set, and $\min_\xi$ represents the minimum value in the set.

$$\xi_{n_j}^{o-nor} = \frac{\max_\xi - \xi_{n_j}^o}{\max_\xi - \min_\xi} \tag{5}$$

By calculating the credibility of the positive symptoms, the credibility of the latest positive symptoms can be obtained as formula (6). Among them, $N_{s_{n_j}^o}$ represents the set of all faulty nodes connected to $S_{n_j}^o$ in the fault propagation model. $| * |$ is used to calculate the number of elements contained in the set.

$$p^{opt}\left(s_{n_j}^o\right) = \frac{\sum_{n_j \in N_{s_{n_j}^o}} \xi_{n_j}^{o-nor}}{\left| N_{s_{n_j}^o} \right|} \tag{6}$$

3.2 Failure Sequencing

Generally speaking, the greater the centrality of a network node and the number of links, the more times the node will be used by various services. Therefore, if an important node fails, there will be more negative symptoms. If an important node fails, but the number of related negative symptoms is small, it means that some negative symptoms are missing or the failure is a false failure. Based on this, the failure model can be optimized based on the importance of the node and the number of negative symptoms of related services.

Through the analysis of the relevant characteristics of the network nodes, it is known that the main factors related to the number of bearer services on the network nodes are the centrality of the nodes and the number of links. The centrality of a node indicates the degree to which the node is in the center of the network. Generally speaking, the more a network node is in the center of the network, the more likely it is that the node will become a key resource of the network, thereby carrying more services. The number of node links refers to the number of edges of the node. Generally speaking, the greater the number of edges of a node, the greater the possibility of passing through the node, and thus the greater the number of services carried.

Use formula (7) to calculate the centrality of the node, and use formula (8) to calculate the number of links of the node. Among them, $n_r \in \psi(n_j)$ represents the network node collection after removing the network node n_j from the network topology. d_{rj} represents the number of links included in the shortest path between network node n_r and network node n_j. $e \in E(n_j)$ represents the edge of network node n_j, and $bw(e)$ represents the bandwidth value of the edge. It can be seen from formula (7) that the larger the value of $CC(n_j)$, the more likely the network node n_j is to be in the center of the network topology, and the greater the number of services carried on it. It can be seen from formula (8) that the larger the value of $DC(n_j)$, the more resources of network node n_j, and the greater the number of services carried on it. Through the normalized formula (9), the total resource Ω_{n_j} of the network node n_j can be obtained.

$$CC(n_j) = \frac{1}{\sum_{n_r \in \psi(n_j)} d_{rj}} \tag{7}$$

$$DC(n_j) = \sum_{e \in E(n_j)} bw(e) \tag{8}$$

$$\Omega_{n_j} = \frac{CC(n_j) - \min_{CC}}{\max_{CC} - \min_{CC}} + \frac{DC(n_j) - \min_{DC}}{\max_{DC} - \min_{DC}} \tag{9}$$

The proportion of the resources of the network node n_j in the total resources is shown in formula (10). Among them, $n_i \in N_s$ represents all network nodes in the network topology. The proportion of negative symptoms related to network node n_j in the total negative symptoms is shown in formula (11). Among them, $S_{n_j}^-$ represents the negative symptoms related to network node n_j, and S^- represents the set of negative symptoms of all network nodes.

$$\eta_{n_j} = \frac{\Omega_{n_j}}{\sum_{n_i \in N_s} \Omega_{n_i}} \tag{10}$$

$$\sigma_{n_j} = \frac{\left| S_{n_j}^- \right|}{\left| S^- \right|} \tag{11}$$

Based on the above analysis, formula (12) can be used to calculate the fault credibility β_{n_j} of network node n_j. Among them, β_{n_j} represents the reliability of the fault. The closer the value is to 1, the higher the reliability.

$$\beta_{n_j} = \frac{\eta_{n_j}}{\sigma_{n_j}} \tag{12}$$

4   Algorithm Description

The fault diagnosis algorithm based on service characteristics under software defined network slicing (FDAoSC) proposed in this paper includes three processes: building an initial fault propagation model, optimizing fault propagation model, and fault set diagnosis. In the step of constructing the initial fault propagation model, the underlying network service provider first receives a collection of abnormal services from the virtual network, and then builds a fault propagation model of the underlying network and abnormal services based on the mapping relationship and the virtual network resources occupied by the abnormal services. In the optimization step of the fault propagation model, the positive symptom values in the fault propagation model are first optimized, and then the fault credibility of the network nodes is calculated and arranged in descending order. In the fault set diagnosis step, first use the sorted fault credibility set to construct a suspected fault propagation model, and secondly determine the fault set based on the maximum coverage algorithm.

The two steps of constructing the initial fault propagation model and optimizing the fault propagation model have been described in the previous section. The following describes the diagnosis steps of the fault collection. Fault set diagnosis includes two subprocesses: constructing fault set and selecting faulty node based on maximum coverage. When constructing the fault set, assume that the number of simultaneous faults is k. Therefore, a set of k suspected faults consisting of 1 to k faulty nodes is constructed at the same time. The set of k suspected failures is denoted by $H = \{n_1, n_2, \ldots, n_k\}$, where the suspected failure nodes are the k suspected failure nodes ranked in the front of the failure credibility set.

In order to evaluate the explanation ability of the suspected fault set $H = \{n_1, n_2, \ldots, n_k\}$ for abnormal services, the explanation ability of the fault set is defined as $EXP(H,S)$, and the calculation is carried out using formula (13). $\prod_{n_j \in H} P(n_j)$ represents the probability that the failures of the suspected failure set $H = \{n_1, n_2, \ldots, n_k\}$ are all failures. $\prod_{s_i \in S} \left(1 - \prod_{n_j \in H} \left(1 - P(s_i|n_j)\right)\right)$ means that any symptom $s_i \in S$ can find at least one failed network node in the suspected failure set $H = \{n_1, n_2, \ldots, n_k\}$. By calculating the k suspected failure sets, the suspected failure set $H = \{n_1, n_2, \ldots, n_k\}$ with the largest value of $EXP(H,S)$ is taken as the final failure set.

$$EXP(H, S) = \prod_{n_j \in H} P(n_j) \times \prod_{s_i \in S} \left(1 - \prod_{n_j \in H} \left(1 - P(s_i|n_j)\right)\right) \qquad (13)$$

5   Performance Analysis

5.1   Network Environment

The network topology in the experimental environment is generated using GT-ITM [7] tool. The generated network topology includes two types: virtual network and underlying network. The network nodes of the virtual network obey the uniform distribution of (5, 15), and the network nodes of the underlying network increase from 100 to 500. The resource allocation from the virtual network to the underlying network uses the classic algorithm [8].

In terms of network node failure simulation, the prior failure probability of the bottom node is set to obey the uniform distribution of [0.001, 0.01]. The state of the network node is updated once at an interval of 20 s to simulate the dynamic characteristics of the underlying network. In terms of algorithm comparison, compare the FDAoSC algorithm in this paper with the fault diagnosis algorithm based on resource bearing relationship (FDAoRBR). Among them, the algorithm FDAoRBR establishes a fault diagnosis model by analyzing the relationship between resource bearers and associating service status with underlying network resources. In terms of algorithm comparison indicators, the analysis is carried out from three aspects: the accuracy rate of fault diagnosis, the false alarm rate, and the diagnosis time. Accuracy refers to the proportion of identified faults in real faults. The larger the value, the more faults identified by algorithm diagnosis. The false alarm rate refers to the proportion of the identified false faults in the identified faults. The smaller the value, the higher the authenticity of the faults diagnosed by the algorithm. Diagnosis duration refers to the duration of the algorithm from receiving the service status to outputting the fault set. The smaller the value, the better the algorithm performance.

5.2 Performance Comparison

The comparison result of the accuracy of the fault diagnosis algorithm is shown in Fig. 2. From the figure, it can be seen that the network size has a small effect on the diagnosis accuracy of the two algorithms, and the accuracy of the algorithm in this paper is higher. It shows that this paper optimizes the symptom set and the fault set, thereby improving the accuracy of the fault diagnosis model.

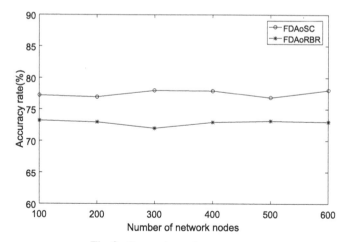

Fig. 2. Comparison of accuracy rate

The comparison result of fault diagnosis false alarm rate is shown in Fig. 3. It can be seen from the figure that the two algorithms have achieved relatively stable diagnosis results under different network scales. The false alarm rate of the algorithm in this paper is lower than that of the traditional algorithm. This is because the algorithm in this paper evaluates characteristics such as symptom status and fault credibility, and better solves the problem of inaccurate fault propagation model caused by dynamic environment. The algorithm in this paper improves the accuracy of the fault diagnosis model data, thereby reducing the false alarm rate.

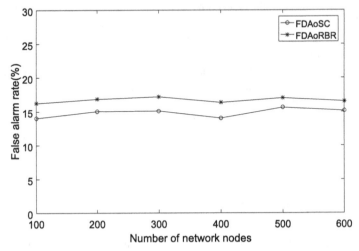

Fig. 3. Comparison of false alarm rate

The result of the comparison of the fault diagnosis duration is shown in Fig. 4. It can be seen from the figure that as the network size increases, the diagnosis time of the two algorithms increases faster. Compared with the existing algorithm, the diagnosis time of the algorithm in this paper has increased slightly. This shows that the algorithm in this paper optimizes faults and symptoms, and there is a certain time overhead. However, the algorithm in this paper constructs a set of suspected faults based on fault credibility, and the time to infer the fault is shorter.

6   Conclusion

The software defined network slicing technology makes the relationship between faults and symptoms in the fault diagnosis algorithm more complicated, resulting in a decrease in the accuracy of the fault diagnosis model and the performance of the fault diagnosis algorithm. To solve this problem, this paper proposes a fault diagnosis algorithm based on service characteristics under software defined network slicing. The algorithm includes three processes: constructing an initial model of fault propagation, optimizing the fault propagation model, and fault diagnosis. The algorithm receives a collection of abnormal services from the virtual network. Based on the mapping relationship and the virtual network resources occupied by abnormal services, the fault propagation model of the underlying network and abnormal services is constructed. The positive symptom

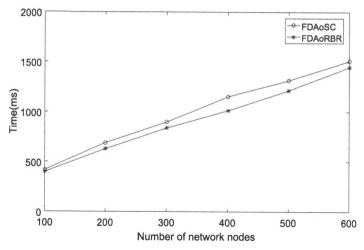

Fig. 4. Comparison of time

value in the fault propagation model of the algorithm team is optimized to improve the fault credibility of the network node. The algorithm uses the sorted fault credibility set to construct a suspected fault propagation model, and uses the maximum coverage algorithm to determine the fault set. The algorithm optimizes the symptom value in the fault propagation model based on the substitutability of the node, establishes a mathematical model of the importance of the node and the number of symptoms of the fault, and optimizes the fault propagation model. The experimental part verifies that the algorithm in this paper improves the accuracy of fault diagnosis.

When evaluating the importance of nodes, this article only considers the centrality of nodes and the number of links. The next step is to enrich this content and analyze the node attributes such as the approximate ideal ranking method and the analytic hierarchy process to better improve the algorithm performance.

Acknowledgement. This work is supported by science and technology project from State Grid Jiangsu Electric Power Co., Ltd: "Technology Research for High-efficiency and Intelligent Cooperative Wide-area Power Data Communication Networks (SGJSXT00DDJS1900168)".

References

1. Kaizhi, H., Qirun, P., Quan, Y., et al.: A virtual node migration method for side channel risk perception. J. Electron. Inf. Sci. **41**(9), 2164–2171 (2019)
2. Wu, B., Ho, P.H., Tapolcai, J., et al.: Optimal allocation of monitoring trails for fast SRLG failure localization in all-optical networks. In: Proceedings of 2010 IEEE Global Telecommunications Conference, Miami, USA, pp. 1–5 (2010)
3. Dusia, A., Sethi, A.S.: Recent advances in fault localization in computer networks. IEEE Commun. Surv. Tutor. **18**(4), 3030–3051 (2016)
4. Rish, I., Brodie, M., Ma, S., et al.: Adaptive diagnosis in distributed systems. IEEE Trans. Neural Netw. **16**(5), 1088–1109 (2005)

5. Jin, R., Wang, B., Wei, W., et al.: Detecting node failures in mobile wireless networks: a probabilistic approach. IEEE Trans. Mob. Comput. **15**(7), 1647–1660 (2016)
6. Yu, Y., et al.: Falcon: differential fault localization for SDN control plane. Comput. Netw. **162**(106851), 1–15 (2019)
7. Zegura, E.W., Calvert, K.L., Bhattacharjee, S.: How to model an internetwork. In: Proceedings of IEEE INFOCOM (1996)
8. Yu, M., Yi, Y., Rexford, J., Chiang, M.: Rethinking virtual network embedding: substrate support for path splitting and migration. ACM SIGCOMM CCR **38**(2), 17–29 (2008)

Security Situation Awareness and Interference Control Method for Power Wireless Private Networks Based on Dynamic Baseline

Jin Huang[1], Weiwei Miao[1], Junzhong Yang[2], Xinglong Wang[2], Linshan Shi[3], Zhengyuan Liu[4], and Peng Yu[4(✉)]

[1] State Grid Jiangsu Electric Power Co., Ltd., Information & Telecommunication Branch, Nanjing 210024, China
[2] State Grid Jiangsu Electric Power Co., Ltd., Taizhou Power Supply Branch, Taizhou 225307, Jiangsu, China
[3] State Grid Chongqing Electric Power Company, Chongqing 404100, China
[4] Beijing University of Posts and Telecommunications, Beijing 100876, China
yupeng@bupt.edu.cn

Abstract. The interference of the power wireless private network will directly affect the service quality of the power business, and then affect the stable operation of the power distribution network. Therefore, it is of great value to study the security situation awareness and interference control technology of electric power wireless private network. Existing research has seldom considered security situational awareness and control methods for power wireless private networks. Firstly, according to the signal-to-noise ratio data obtained by the network management system, combined with the upper and lower baselines and user proportion, the interference is identified. According to the interference scenario, the corresponding power adjustment scheme is proposed according to the different scale of users to ensure the stability of the system. In view of the influence of power adjustment, the dynamic change method of interference identification baseline is proposed, which provides reasonable interference control requirement standard for base station to adjust transmission power control interference, and improves the flexibility of service performance guarantee.

Keywords: Power wireless private network · Interference monitoring · Interference identification · Interference control

1 Introduction

With the widespread application of power wireless private network, convenient access methods are provided for control services such as power distribution automation, source-network load-storage interaction, and management services such as electricity information collection, mobile operations, video surveillance and so on [1]. The power wireless private network inherits the advantages of wireless network, such as flexible networking, convenient construction, and mature application. At the same time, the dedicating of

M. Cheng et al. (Eds.): SmartGIFT 2020, LNICST 373, pp. 77–92, 2021.
https://doi.org/10.1007/978-3-030-73562-3_7

frequency band, equipment, and network avoiding the limitations of the wireless public network in terms of bandwidth, delay, service interruption rate, safety and reliability. It can effectively supplement the wired transmission network and efficiently solve the "last mile" access problem of power communication, open up the "nerve endings" of the power communication network, and have the incomparable advantages of traditional wired communication and wireless public network communication [2].

However, with the rapid increase of the number of power wireless private network access terminals, in the scenario of unified access of various mixed services such as source network load storage, distribution monitoring, mobile applications, etc., the current centralized processing method through the core network can not meet the real-time requirements of load control power services. At the same time, power wireless private network also inherits the wireless system's channel opening, network sharing, terminal mobility and other characteristics, which also poses a huge challenge to the security of power services. Since power wireless private network carries a large number of electric power dedicated services, it is easy to cause service transmission interruption under the condition of wireless interference, which will affect the stable operation of the power system. Therefore, in view of the communication architecture of "terminal base station core network master station" of power wireless private network, this paper carries out security situation awareness on the electromagnetic space of power wireless private network and analyzes the potential interference risk, it is very important to monitor the flow and service quality of power wireless private network terminal and control the interference of wireless private networks for power systems [3, 4].

The existing research on interference of power wireless private network mainly includes two aspects: internal interference control and external interference control. For internal interference in the system, interference coordination and power control are often used to realize the control. For the external interference sources, the data analysis and other devices are used to locate the interference source. However, there are fewer studies on interference threshold control methods. When the interference is found, how to optimize the specific interference control parameters and minimize the influence of the network is also lack of limited analysis [5, 6] (Fig. 1).

Based on the above analysis, this paper proposes a dynamic baseline-based power wireless private network security situation awareness and interference control method. In this method, the interference monitoring and identification method is firstly proposed, and then the interference control method based on power adjustment is proposed. Finally, a dynamic adjustment method for the interference identification baseline is given. Based on this method, only through the scheduling of the base station side, the interference control of the network can be realized and the service quality of the service can be improved.

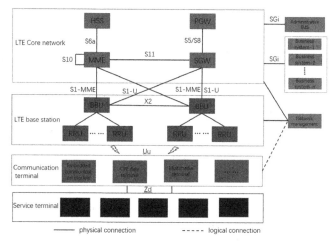

Fig. 1. LTE wireless private network architecture.

2 Research Background and Related Work

In the paper [3], after analyzing the influence of strong interference sources outside the power wireless private network, in view of the problem that the interference source in the power wireless private network cannot be confirmed, a fault tracking and positioning method for LTE power wireless private network external interference is proposed. The method first determines the range of the interfered base station, and then confirms the search range of the interference source; performs 3D scene modeling for the search range, then planning the Drive Test (DT) path for drive test, extracting the data characteristics of the Drive Test, analyzing and approaching interference area; finally, through forward ray tracing, the actual interference source location is evaluated and screened out.

The paper [5] draws lessons from the public network operation and maintenance standards to establish interference judgment indicators. It is considered that higher than -120 dBm means interference exists, higher than -110 dBm means that the interference begins to worsen various indicators, higher than -105 dBm means that the interference is more serious. The system external disturbances such as blocking interference, intermodulation interference, stray interference and LTE network interference are analyzed respectively. Finally, the confirmation methods and interference treatment methods for these interferences are given.

This paper [6] designs a progressive, step-by-step measurement and step-by-step approach interference location method based on the principle of radio direction finding and manual interference checking and positioning process. Firstly, the interference source search range is determined according to the disturbed cell, equipment terminal, frequency sweep/Drive Test point, and user complaint point in the network; then, the interference source search range is modeled in three-dimensional scene; then, the iterative search mechanism is used to plan the Drive Test path and implement the Drive Test, extract and analyze the characteristics of the Drive Test data, and use multiple heuristic rules to gradually compress and approach the interference source; finally, when the interference source search area is compressed enough, the beam tracking wireless propagation model is used to quickly evaluate the candidate interference source location, and the interference source location results are given.

The paper [7] firstly analyzes the broadband interference of the 230 MHz full frequency band and the narrowband interference of high-power data transmission station. According to the distribution of the interference frequency and the impact on the service, it can be divided into three levels: broadband full-frequency interference (high), narrowband co-channel interference (medium) and narrow-band adjacent frequency interference (low). The causes and effects of broadband full frequency interference and narrowband co-frequency interference are analyzed. For broadband full frequency interference, due to the use of dual-antenna technology in 230 MHz power wireless private network, the UHF filter bank interference suppression algorithm is used in the uplink to suppress the interference, and the user-level power control algorithm is used in the downlink to increase the signal strength of specific users. For narrowband co-channel interference, random and fast frequency hopping methods are used to avoid narrowband co-channel interference.

The paper [8] is oriented to the dense urban environment and the network coverage area is divided into multiple geographical scenes by using spatial clustering technology. According to the terrain and feature information provided by the three-dimensional electronic map, the scene-oriented ray tracing propagation model is used to analyze the outdoor wireless signal coverage distribution; in order to solve the analysis error caused by the uncertainty of model parameters, a scene-oriented propagation model parameter correction is proposed. By using the Drive Test data, the multi-objective optimization model of propagation model parameters for scene is established, and multi-objective evolutionary algorithm is used to optimize the direct radiation, reflection, and diffraction coefficients of the model.

The paper [9] introduces dynamic spectrum sensing and spectrum planning and management schemes in LTE230 system, and proposes a sensing-based interference avoidance method. The authorized frequency points are separated to a certain extent by the system, that is, the 230 MHz frequency band is divided into multiple resource groups, and multiple systems conduct spectrum sensing based on resource groups to obtain available resource group resources and realize resource sharing. In order to ensure the normal operation of the sensing system and high real-time and high reliability service transmission, new authorized frequency points are redistributed within the limited range of each resource group. Finally, tests were carried out for co-channel interference and adjacent channel interference.

This paper [10] mainly analyses the situation that the power wireless private network developed on 230 MHz spectrum resources may share the same frequency band with 230 MHz data transceivers in the application process. In order to ensure that the two systems operate stably and efficiently at the same time, a trust-based cooperative spectrum sensing algorithm is proposed. According to the user behavior characteristics, the users are divided into trusted users, invalid users and interfering users, and corresponding judgment methods are given to untrusted users. At the same time, the relationship between the local detection results of the trusted users and the users participating in the cooperation and the signal-to-noise ratio of the communication channel are analyzed, and then the weight of each user's credibility is defined. On this basis, a cooperative spectrum sensing algorithm based on trust degree is proposed. The simulation results show that compared with the traditional hard-decision cooperative spectrum detection method, the cooperative sensing algorithm based on sensing user trust can obtain higher detection performance.

Although the existing schemes can monitor, identify and control the interference to a certain extent, there are still some problems. (1) Additional hardware and software are required; (2) The threshold for interference determination is relatively fixed, and there is a lack of consideration of the impact after dynamic control; (3) The specific control method for the interference frequency band is not considered; (4) The identification method of interference is not considered, and the requirements for the base station are very high, so some additional hardware modules need to be upgraded. The method proposed in this paper is mainly based on the network management system data to realize the effective judgment of the existence of interference, and through the power adjustment and the adjustment of the corresponding interference baseline, to achieve the effective control of the terminal interference, without additional hardware support. At the same time, it has both flexibility and effectiveness, which can effectively improve service transmission efficiency and ensure the reliability of power communication systems.

3   Security Situation Awareness and Control Method Based on Heuristic Algorithm

The process of the interference control mechanism of the LTE power wireless private network mainly includes three stages of monitoring, analysis and adjustment. The flow chart is shown below, the three processes form a closed loop process (Fig. 2).

A detailed description of each of the above stages is as follows.

3.1   Interference Signal Analysis Based on Network Monitoring

The monitoring stage is mainly to obtain the required network system parameter information through various methods, and calculate the terminal service quality. The data can be obtained from the interface of network management system and LTE power wireless private network terminal. The data and definitions that need to be obtained include:

$P_i^l(t)$: The transmission power of the i-th base station on the l-th subcarrier at time t.

$G_{ij}^l(t)$: The channel gain from the i-th base station to the j-th user's l-th subcarrier at time t.

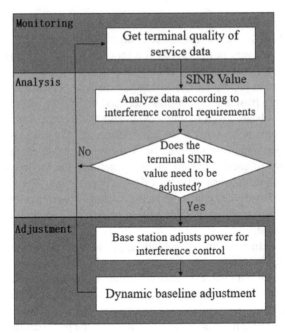

Fig. 2. Analysis of interference control flow.

$N_0(t)$: Noise power spectral density at time t.

W: Self-channel bandwidth.

$SINR_{ij}^l(t)$: The signal-to-noise ratio of the l-th subcarrier from the i-th base station to the j-th user at time t. The calculation formula is:

$$SINR_{ij}^l(t) = \frac{P_i^l(t)G_{ij}^l(t)}{\sum\limits_{j\in U} P_i^l(t)G_{ij}^l(t) + N_0(t)W} \tag{1}$$

$SINR_{ij}(t)$: The signal-to-noise ratio of the j-th user. The calculation formula is:

$$SINR_{ij}(t) = \sum\limits_{l\in L} SINR_{ij}^l(t) \tag{2}$$

Where L is the set of subcarriers.

After obtaining these data through the network management system, it enters the analysis stage.

3.2 Analysis of Cyberspace Interference Situation Based on Baseline

In the analysis stage, by comparing the terminal signal-to-noise ratio data obtained in the monitoring stage with the preset baseline, the initial upper and lower baselines of acceptable SINR are set to γ_{max} and γ_{min} respectively, and the base station i is judged. Whether the served terminal has strong interference, the identification index $f_i(t)$ is defined as follows:

$$f_i(t) = Card\{i|SINR_{ij}(t) < \gamma_{\min}\}/|U_i| \tag{3}$$

The $card\{\cdot\}$ part represents the number of base station i whose signal-to-noise ratio is not higher than the minimum threshold, and U_i is the number of users served by base station i. $f_i(t)$ shows the proportion of users whose current signal-to-noise ratio is affected. According to this proportion, the process of interference identification technology to be adopted is as follows:

Step 1: Calculate the proportional value $f_i(t)$ for each base station i. If $f_i(t) = 0$, it is judged that there is no interference and no control will be given and return to the monitoring stage; if $0 < f_i(t) < \beta$, β is the proportion threshold, it is considered that the number of terminals subject to interference is small, and then proceed to step 2; if $\beta < f_i(t) < 1$, it is considered that a large number of terminals are interfered, and proceed to step 3.

Step 2: Determine the interfered terminal set U'_i and the corresponding subcarrier set L'_i through the analysis of the network management system, determine the base station transmission power P_i and the maximum transmission power P_{max}, and enter the adjustment stage.

Step 3: Analyze the network management system to determine the interfered terminal set U'_i, determine the base station's transmission power P_i and the maximum transmission power P_{max}, and enter the adjustment stage.

3.3 Dynamic Baseline Adjustment Based on Long Short Memory Model

According to interference control requirements, for terminals whose signal-to-noise ratio is based on the system baseline value, the base station accessed by the terminal during the adjustment phase needs to adjust its transmit power to the connected user terminal to perform interference control to improve terminal service performance. The specific steps are as follows:

1) For the scenario of $0 < f_i(t) < \beta$, the power P_i^l of the subcarrier set L_i' is increased according to a uniform step size Δd, make $P_i \leftarrow P_i + |U_i'|\Delta d$; for the scenario of $0 < f_i(t) < \beta$, make $P_i \leftarrow P_i + \Delta e(\Delta e > \Delta d)$;

2) Determine whether P_i is less than P_{max}. If it meets the requirements, determine whether the user SINR in the adjusted terminal set U_i' are all greater than γ_{min} and less than γ_{max}. If so, the adjustment ends, the dynamic baseline update is performed, and return to the monitoring stage; otherwise return to step 1), if the power constraint is not satisfied, go to step 3;

3) Report the interference alarm to the network management system, after the adjustment is completed, return to the monitoring stage.

In step 2), the dynamic baseline update process is as follows:

The baseline γ_{max} and γ_{min} are used to indicate the relatively stable value of the SINR value over a period of time. The determination of the baseline is a key issue: if the baseline is too strict, the base station will frequently adjust the power; on the contrary, if the baseline is set too wide, the related service quality degradation cannot be dealt with in time, resulting in poor reception of the user terminal. The baseline calculation algorithm should be simple to implement, but with a high degree of confidence, in order to well reflect abnormal changes in business performance, that is, the extent to which the terminal service quality is affected by signal interference.

We set the baseline value of the acceptable signal-to-noise ratio SINR of the terminal in the system. This project will use the dynamic baseline algorithm, that is, after the power is adjusted, the corresponding baseline will be increased accordingly. With real-time monitoring of terminal service quality and timely detection of business performance abnormalities and interference control as the goal, this report adopts a dynamic method and the specific algorithm is described as follows:

1) Initialize the static baseline values γ_{max} and γ_{min};

2) For the relevant user j after the adjustment, it is assumed that a total of N SINR data are obtained, which are respectively recorded as $X_1, X_2, X_3, ...,X_i, ...,X_N$, suppose there are N' data of terminal SINR that are more severely interfered ($X_i < \gamma_{min}$), calculate the average value $\overline{X} = \frac{1}{N'} \sum_{i=1}^{N} (X_i - \gamma_{min}) \ (X_i < \gamma_{min})$ of the data below the static baseline value.

3) For the current SINR of the interfered terminal, the SINR set $D'(t) = \{D_1(t), D_2(t), ...,D_m(t)\}$ for different terminals i in space at time t, the goal is to predict the next parameter $D(t + 1)$. The core of LSTM is the hidden layer sequence $h = \{h1, h2, ..., h_n\}$, the output is the predicted time series value $Z_t = \{z_1(t), z_2(t), ..., z_n(t)\}$, its calculation method as follows:

$$z_t = W_{hz}h_t + b_y \tag{4}$$

The calculation process of the hidden layer sequence h_t at time t is as follows:

$$h_t = o_t \tanh(c_t) \tag{5}$$

$$o_t = \sigma(W_{Lo}L(t) + W_{ho}h_{t-1} + W_{co}c_t + b_o) \tag{6}$$

$$c_t = f_t c_{t-1} + i_t \tanh(W_{Lc}L(t) + W_{hc}h_{t-1} + b_c) \tag{7}$$

$$f_t = \sigma(W_{Lf}L(t) + W_{hf}h_{t-1} + W_{cf}c_{t-1} + b_f) \tag{8}$$

$$i_t = \sigma(W_{Li}L(t) + W_{hi}h_{t-1} + W_{ci}c_{t-1} + b_i) \tag{9}$$

Where W is the weight coefficient matrix of each layer, b is the bias vector of each layer, i, f, c, and o are input gate, forget gate, cell state, and output gate respectively; σ and tanh are sigmoid and hyperbolic tangent, respectively activation function.

In order to ensure the accuracy of the model, the training process of the LSTM model is as follows:

Step 1: Calculate the output value of LSTM cell according to (4)–(9);

Step 2: Calculate the error term of each LSTM cell, including two back propagation directions according to time and network level;

Step 3: Calculate the gradient of each weight according to the corresponding error term;

Step 4: Apply gradient-based optimization algorithm to update weights.

In view of the above training process, this project will compare and analyze different hidden layer cell structures and different gradient optimization algorithms, and analyze the rationality of the prediction results. At the same time, compare the LSTM prediction model of this project with other neural network algorithms, S-ARIMA and other time series algorithms to verify its accuracy and effectiveness, so as to obtain the most accurate observation parameter results.

4) The baseline value is adjusted to $\gamma'_{min} = \gamma_{min} + \frac{1}{2}(\overline{X} + \sum_i \frac{z_i(t)}{|z(t)|})$, then enter the next stage for interference control and enter the monitoring stage.

4 Simulation and Implementation Analysis

In this chapter, the security situation awareness and interference control method proposed in the previous chapter will be verified by simulation, and the performance of the algorithm will be analyzed by comparing with other algorithms. The experiment is completed in the MATLAB simulation platform.

4.1 Simulation Environment Settings

In the simulation scenario, the network area is 1 000 m × 1 000 m, including 19 macro base stations with 57 cells and 80 terminals which are randomly deployed in the network (Fig. 3).

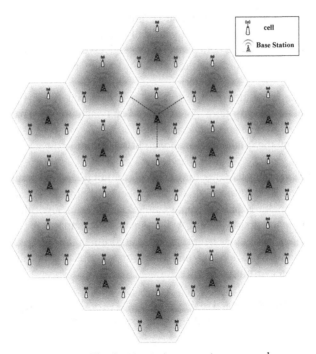

Fig. 3. Simulation scenarios.

The parameter values used in the simulation process are shown in Table 1 below.

Table 1. Simulation parameters

Parameters	Value	Unit
Antenna gain	10	dBi
Road loss model	$135.88 + 37.6\log_{10}d$	dB
power spectral density	-118.4	dBm
bandwidth	180	kHz
Speed requirement	0.5–1.5	Mb/s
RB amount	80	-
β	0.9	-
Maximum base station power	500	W
Δd	0.1	dBm
Δe	0.2	dBm
$P_{user-min}$	-120	dBm
γ_{min}	-10	dB
γ_{max}	60	dB

Based on the simulation parameters given above, the simulation results of the distributed algorithm are given below.

4.2 Analysis of Simulation Results

With the simulation analysis, the interval distribution of SINR get by terminals with different signal is shown in the figure below. We can find that BS number 13 and 18 are affected with interferences (Fig. 4).

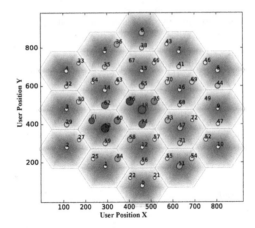

Fig. 4. Distribution interval diagram of signal strength.

Next, we use the time-varying noise as the experimental environment, as shown in Fig. 5 below, respectively set up three groups of simulation experiments, corresponding to high, medium and low noise environment. We find that SINR decreases with the increase of noise, so we need a control method to maintain network service quality.

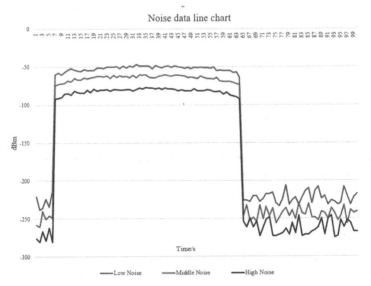

Fig. 5. Noise environment change chart.

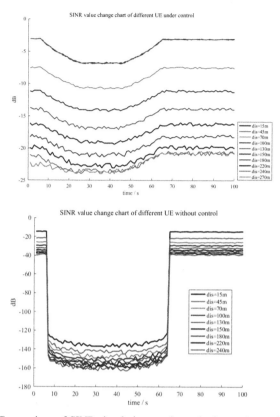

Fig. 6. Comparison of SINR simulation results under low noise environment.

Figure 6, Fig. 7 and Fig. 8 show the comparison of SINR values of several UE devices under controlled and uncontrolled conditions when the ambient noise is low, medium and high. It can be seen that no matter the environmental noise is high, medium and low, the fluctuation of SINR value in the controlled situation is much less than that in the uncontrolled situation, and the network also has better quality of service. Therefore, this algorithm can better maintain the network quality of service at a better level by dynamically adjusting the power of the base station in the case of large environmental noise.

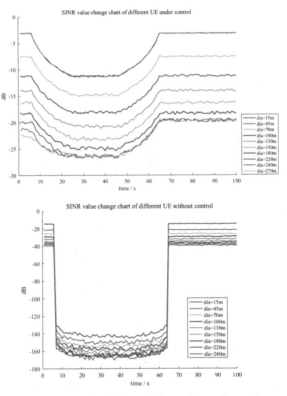

Fig. 7. Comparison of SINR simulation results under medium noise environment.

5 Conclusion

After fully investigating the existing network interference detection, interference analysis and interference control algorithms, this paper presents the advantages and disadvantages of each algorithm for different algorithms. At the same time, this paper proposes a dynamic baseline based security situation awareness and interference control algorithm for power wireless private network, and builds an experimental simulation and Simulation for the algorithm using MATLAB software The real results show that the continuous enhancement of noise will continue to reduce the network quality of service and the SINR value will continue to decrease. That is, the algorithm can maintain the SINR of all user devices to a certain extent.

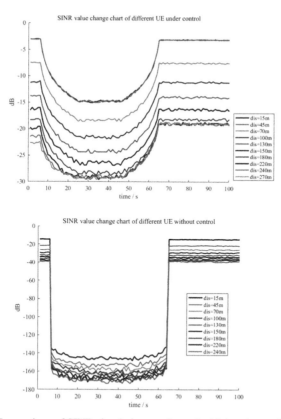

Fig. 8. Comparison of SINR simulation results under high noise environment.

Acknowledgement. This work is supported by Science and Technology Project from of State Grid Corporation of China: "Key technology Research and Application for Ubiquitous Inter-networking Security for Power wireless private network services (5700-201918229A-0-0-00)".

References

1. Cao, J., Liu, J., Li, X.: A power wireless broadband technology scheme for smart power distribution and utilization networks. Autom. Electric Power Syst. **37**(11), 76–80 (2013)
2. Sun, S., Cheng, Y.: Research on LTE power wireless private network for service coverage. Power Inf. Commun. Technol. **13**(4), 6–10 (2015)
3. Wang, H., Gu, S., Wanyan, S., et al.: Tracking and locating method of interference source in power wireless private network. Guangdong Electric Power **32**(6), 86–93 (2019)
4. Liu, R., Yu, J., Zhao, G., et al.: High reliability planning method for power wireless private network with low self interference. Power Syst. Autom. **42**(17), 162–167 (2018)
5. Kong, W., Luo, X., Liu, Y., et al.: Interference analysis of power wireless private network. Commun. Technol. **52**(01), 122–128 (2019)
6. Yin, J.: Research and application of location technology of unknown interference source outside LTE power wireless private network. Beijing University of Posts and Telecommunications (2019)

7. Tong, J., You, Q., Sun, C., et al.: Research on interference avoidance strategy of 230 MHz power wireless private network. Power Inf. Commun. Technol. **18**(3), 27–33 (2020)
8. Teng, J.: Research and application of high precision coverage interference sub technology in TD-LTE power wireless private network. Beijing University of Posts and Telecommunications (2019)
9. Zheng, W., Fang, J., Chen, P., et al.: Implementation and verification of an automatic interference avoidance method based on dynamic sensing. In: Proceedings of the Third Smart Grid Conference, pp. 507–513+520 (2018)
10. Cao, J., Li, W., Zhang, Y.: A trust based cooperative spectrum sensing algorithm for power wireless private networks. Telecommun. Sci. **31**(3), 136–141 (2015)

Energy-Aware Blockchain Resource Allocation Algorithm with Deep Reinforcement Learning for Trusted Authentication

Lifang Gao[1], Xiaotao Zhang[2], Tingfeng Liu[2], Huifeng Yang[1], Boxian Liao[3(✉)], and Jing Guo[2]

[1] State Grid Hebei Information & Telecommunication Branch, Shijiazhuang, China
[2] Aostar Information Technologies Co., Ltd. 2688 Xiyuan Dadu, South of Modern Industrial Port, Chengdu, China
[3] State Key Laboratory of Networking and Switching Technology, Beijing University of Posts and Telecommunications, Beijing, China
boxian_liao@bupt.edu.cn

Abstract. Internet of things (IoT) technology is in continuous development, and the access of the IoT power terminal is facing various security threats such as data tampering and malicious attacks. Thus, we propose a blockchain-based edge-terminal collaborative resource allocation architecture to solve these security problems, which places the terminal trusted authentication data on the blockchain to realize the security of the terminal authentication information. Since the mining process of the blockchain system will generate a large number of computing intensive tasks, this paper establishes an energy-oriented blockchain mining task offloading model, and proposes the energy-aware blockchain resource allocation (EABRA) algorithm with deep reinforcement learning (DRL) to jointly optimize the offloading decision and transmission power allocation decision. Finally, the simulation results show that the EABRA algorithm can save 68.87% energy consumption than the Random algorithm, which verifies the correctness and feasibility of the scheme.

Keywords: Internet of things · Edge-terminal collaborative · Trusted authentication · Task offloading

1 Introduction

Recently, the power terminals of the Internet of things (IoT) present a variety of characteristics, and most power terminals are more vulnerable to be attacked due to their low security. The traditional terminal authentication mechanism faces many security problems such as data leakage and data tampering. And the centralized authority solution is not suitable for the authentication and access of highly distributed IoT devices [1]. The distributed characteristics of power terminals need a self-protection mechanism that does not depend on the central authority [2]. In order to realize the trusted authentication

M. Cheng et al. (Eds.): SmartGIFT 2020, LNICST 373, pp. 93–103, 2021.
https://doi.org/10.1007/978-3-030-73562-3_8

of power terminal and reduce the security risk of terminal access, we use the security and tamper-resistant of blockchain technology, and use mobile intelligent terminal as the key carrier of identity authentication and electronic signature. The core data is placed on the blockchain, the blockchain is used to realize the safe storage of authentication information, and the smart contract is used to realize the safe and automated execution of the authentication process [3].

Based on the analysis of the research status at home and abroad, it can be seen that many scholars have made a lot of research achievements on blockchain-based trusted authentication methods. Neisse et al. [4] proposed a blockchain-based platform to enhance the transparency and traceability of network security authentication information and realize the trusted exchange of IoT security authentication information. Cui et al. [5] proposed a blockchain-based multi-sensor network authentication scheme for the IoT, which builds a blockchain network between different types of nodes to achieve mutual authentication of node identities in different communication scenarios. Thakker et al. [6] proposed a data management system based on blockchain technology to ensure the security of access terminals, and the authenticated terminals can ensure data integrity and provide trusted data storage. The non-tamper ability of blockchain can well guarantee the trust of the system. However, the application of blockchain technology is limited by the mining process, that is, miners (mobile terminals) are required to complete computing intensive tasks, which puts forward extremely high requirements for the computing capacity of the terminals [7].

We introduce mobile edge computing (MEC) technology to solve the problem of the limited computing capacity of the blockchain system. In order to achieve the integrity and validity of authentication information, we then propose a blockchain-based edge-terminal collaborative resource allocation architecture to ensure the trust of terminal authentication.

2 System Model

2.1 Application Model

As shown in Fig. 1, the blockchain system is located at the end layer, and we store the trusted authentication information of the terminals in the block. The MEC server is located in the edge layer, and the edge node can be connected with the base station. Let $N = \{1, 2, \ldots, N\}$ represent the set of base stations and $M = \{1, 2, \ldots, M\}$ represent the set of mobile terminals. Suppose that each mobile terminal can perform the mining task K_n of the blockchain system as a miner x_n, and the mobile terminal can access its nearest edge node for collaborative task offloading. We use f_n and f_m (CPU cycles/s) to represent the computing capability of the edge node and the mobile terminal respectively. For miner x_n, we use $<D_n, \tau_n, X_n>$ to denote the data size D_n (bit), the completion time τ_n (second), and the computing intensity X_n (CPU cycles/bit) of the mining task K_n. The symbols used in this paper are summarized in Table 1.

Table 1. Notation definitions.

Symbol	Definition
N	The set of base stations
M	The set of mobile terminals
D_n	The data size of the task
τ_n	The completion time of the task
X_n	The computing intensity of the task
f_m	The computing capability of mobile terminal
f_n	The computing capability of edge node
ζ_m	The computing energy efficiency coefficient of mobile terminal
ζ_n	The computing energy efficiency coefficient of edge node
P_s	The static circuit power
$P_n(t)$	The transmission power between mobile terminal and edge node
$g_n(t)$	The channel gain between mobile terminal and edge node
Γ_{min}	The minimum computing power required by blockchain system

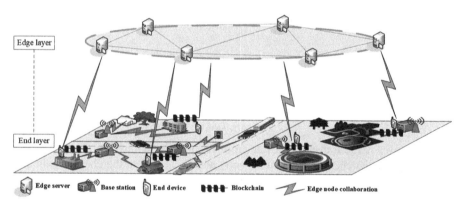

Fig. 1. The blockchain-based edge-terminal collaborative resource allocation architecture.

2.2 MEC Model

In the MEC system, we assume that all mobile terminals and edge nodes have computing capability to perform the mining task of blockchain. Due to the limited computing capability of mobile terminal, it may not be able to handle a large number of computing intensive tasks, so we use MEC server to solve this problem. The MEC server has powerful computing resources. We can offload the computing tasks of the terminal to the edge node for collaborative computing, so as to improve the computing speed of the system. Therefore, we consider two different calculation modes, and let $d_n(t) \in \{0, 1\}$ represent computing offloading decision of miner x_n. $d_n(t) = 0$ indicates that the miner

x_n selects mode 0, that is, the mobile terminal performs computing task. $d_n(t) = 1$ indicates that the miner x_n chooses mode 1, that is, the mobile terminal offloads the computing task to the edge node for computing.

Mobile Terminal Computing

We analyze the performance of miner x_n when performing computing task on the mobile terminal in this mode. The delay of local computing includes the time for miner x_n to complete the task. We use f_m to represent the computing capability of the mobile terminal, that is, the time for miner x_n to perform the mining task K_n can be expressed as:

$$T_m(t) = \frac{D_n X_n}{f_m} \tag{1}$$

Let χ represent the number of CPU cycles required by the mobile terminal to process 1-bit computing task. The computing rate of the mobile terminal in this mode can be denoted as:

$$r_m(t) = \frac{f_m}{\chi} \tag{2}$$

We use ζ_m to represent the computing energy efficiency coefficient of mobile terminal, so the energy consumption of the system can be expressed as:

$$E_m(t) = \zeta_m (f_m)^3 T_m(t) \tag{3}$$

Edge Node Collaborative Computing

The time taken by miner x_n to upload K_n to the edge node can be denoted as:

$$T_n^u(t) = \frac{D_n}{r_n(t)} \tag{4}$$

where $r_n(t)$ is the transmission rate from the mobile terminal to the edge node, we let B denote the channel bandwidth, $g_n(t)$ and $P_n(t)$ denote the channel gain and transmission power between the mobile terminal and the edge node respectively.

$$r_n(t) = B \cdot \log_2(1 + \frac{P_n(t) g_n(t)}{\sigma_n^2(t)}) \tag{5}$$

where $\sigma_n^2(t)$ is the noise power between the mobile terminal and the edge node. We use f_n to represent the computing capability of the edge node, so the time for the edge node to complete the mining task K_n can be expressed as:

$$T_n^e(t) = \frac{D_n X_n}{f_n} \tag{6}$$

We assuming that the number of CPU cycles in the task buffer of the edge server is Q_n, then the queuing delay of miner x_n can be denoted by:

$$T_n^q(t) = \frac{Q_n}{f_n} \tag{7}$$

We use ζ_n to represent the computing energy efficiency coefficient of edge node, and P_s to represent the static circuit power. In this mode, the energy consumption of the system can be denoted as:

$$E_n(t) = P_n(t)T_n^u(t) + \zeta_n(f_n)^3 T_n^e(t) + P_s T_n^q(t) \tag{8}$$

Additionally, we use $E_{tot,n}(t)$ to represent the total energy consumption of the system, which can be denoted as:

$$E_{tot,n}(t) = (1 - d_n(t))E_m(t) + d_n(t)E_n(t) \tag{9}$$

2.3 Blockchain System

We can put the authentication information of mobile terminals in the blockchain system. And we choose some nodes with high voting rate as consensus nodes of blockchain to participate in block generation and verification, which can improve the system performance [8]. In the consensus process of the blockchain system, we use the Delegated Byzantine Fault Tolerance consensus mechanism. The nodes are divided into agent nodes and ordinary nodes. The agent nodes have the right to keep accounts. The ordinary nodes can see the consensus process and synchronize the ledger information. The number of votes for consensus nodes depends on the number of stakes and available computing resources. The available computing resources refer to the remaining computing resources of the node after the offloading task is processed.

We use $\mathbf{S}(t) = \{S_1(t), S_2(t), ..., S_n(t)\}$ and $\boldsymbol{\Gamma}(t) = \{\Gamma_1(t), \Gamma_2(t), ..., \Gamma_n(t)\}$ to represent the set of the stake and available computing resources. Assume that there is a data buffer in the edge server to store offloading tasks that have arrived but have not yet been executed. Additionally, we use Ω_m and Γ_{min} to represent the total computing capability of the edge server and the minimum computing resources required by the blockchain system, respectively. The computing resources available for the blockchain system in the edge server can be denoted as:

$$\Gamma_n(t) = \max\{\Omega_m - \Omega_n(t), \Gamma_{min}\} \tag{10}$$

Let ρ_n to represent the processing density, so the dynamics of the system processing queue can be defined as:

$$\Omega_n(t+1) = \max\{\Omega_n(t) - f_m + \rho_n r_{tot}(t), 0\} \tag{11}$$

where $r_{tot}(t)$ is the total computing rate of the system, and it can be denoted as:

$$r_{tot}(t) = (1 - d_n(t))r_m(t) + d_n(t)r_n(t) \tag{12}$$

3 Problem Formulation

We model the optimization problem as a Markov Decision Process (MDP) to obtain the optimal resource allocation strategy. Due to the dynamic characteristics of the system, we propose an algorithm based on deep reinforcement learning (DRL) asynchronous advantage actor-critic (A3C) to solve this problem. We use tuple $<\mathbf{S}, \mathbf{A}, Pr, r>$ to define the Markov decision process, where $\mathbf{S}$ is the state space, $\mathbf{A}$ is the action space, Pr is the state transition probability, and r is the reward function.

3.1 Optimization Objective

In order to realize the reasonable resource allocation of the system, we propose an optimization problem to minimize the total energy consumption of the system, and optimize the offloading decision and transmission power allocation decision. We express the optimization problem as follows:

$$
\min \sum_{t=0}^{T-1} \sum_{n=1}^{N} E_{tot,n}(t)
$$

$$
\begin{aligned}
\text{s.t.} \quad & T_{tot,n} < \tau_n & \text{C1} \\
& 0 \le P_{tot,n}(t) \le P & \text{C2} \\
& 0 \le f_m < f_n \le f_{max} & \text{C3} \\
& d_n(t) \in \{0, 1\} & \text{C4}
\end{aligned}
\tag{13}
$$

In order to meet the requirements of task delay, constraint C1 ensures that the total time for the system to complete the mining task does not exceed the completion time τ_n. Constraint C2 means that the sum of transmission power does not exceed the total power P. Constraint C3 ensures that all CPU frequencies are non-negative and finite. Constraint C4 ensures that the task offloading decision is effective.

3.2 Problem Transformation

State Space
We express the state space as the combination of channel condition $\mathbf{G}(t) = \{g_n(t), g_{n,k}(t)\}$ and available computing resource $\mathbf{\Gamma}(t) = \{\Gamma_1(t), \Gamma_2(t), ..., \Gamma_n(t)\}$ of the MEC server:

$$
\mathbf{S}(t) \triangleq \{\mathbf{G}(t), \mathbf{\Gamma}(t)\}
\tag{14}
$$

Action Space
We use $A(t) = [\mathbf{d}(t), \mathbf{P}(t)]$ to define the action space. We define the task offloading decision $\mathbf{d}(t)$ and the transmission power allocation decision $\mathbf{P}(t)$ as:

$$
\mathbf{d}(t) \triangleq \{d_1(t), d_2(t), ..., d_N(t)\}
\tag{15}
$$

$$
\mathbf{P}(t) \triangleq \{P_1(t), P_2(t), ..., P_N(t)\}
\tag{16}
$$

State Transition Probability
After performing an action, the probability of leaving the state $s(t)$ to the next state $s(t + 1)$ can be defined as:

$$
Pr(s(t + 1)|s(t), a(t)) = \int_{s^t}^{s^{t+1}} f(s(t), a(t), s)\,ds
\tag{17}
$$

where f is the state transition probability density function.

Reward Function

The reward function in this system can be defined as:

$$r_s = \begin{cases} R(t), & \text{if } C1 - C4 \text{ are satisfied} \\ 0, & \text{otherwise} \end{cases} \tag{18}$$

where $R(t) = \dfrac{1}{\sum\limits_{n=1}^{N} E_{tot,n}(t)}$.

3.3 Problem Solution

A3C is a parallel implementation of deep reinforcement learning asynchronous method [9]. A3C algorithm is to create multiple parallel environments on a machine, put actor and critic in multiple different threads for training and assign tasks, and update the parameters of local network to the global network each single thread completes the learning, and acquire the comprehensive learning of updated parameters from the global network regularly. Each thread will learn from the environment independently and explore different strategies in parallel. Therefore, we use A3C to explore the optimal offloading decision of edge-terminal collaboration.

4 Simulation Results and Analysis

In order to evaluate the performance of the EABRA algorithm under different parameters, we use TensorFlow 2.0 based on Python 3.7 for simulation, and establish a model composed of MEC system and blockchain system, in which the network coverage radius is about 500 m. The main simulation parameters are summarized in Table 2.

Table 2. Simulation parameters.

Parameter	Value
The computing capability of mobile terminal f_m	1 GHz
The computing capability of edge node f_n	2.4 GHz
The processing density χ	737.5 cycles/bit
The noise power density N_0	-174 dBm/Hz
The bandwidth B	180 KHz
The learning rate for actor network η_a	0.001
The learning rate for critic network η_c	0.01
The static circuit power P_s	0.05 W
The data size of the task D_n	0.42 MB

As shown in Fig. 2, we can use TensorBoard, the built-in module of TensorFlow, to see the visualization of A3C algorithm architecture. We can see that the A3C architecture consists of a global network and eight worker agents. The A3C algorithm starts from building a global network.

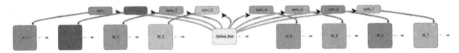

Fig. 2. Visualization of DRL algorithm based on TensorBoard.

Figure 3 shows the internal structure of a worker agent. Each worker agent has actor network and critic network, and can interact with the surrounding environment. After the interaction, worker agents will update the global network parameters according to their own network parameters.

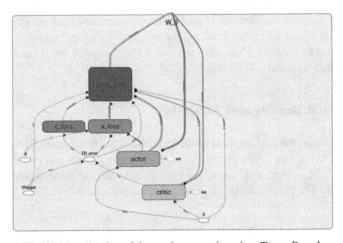

Fig. 3. Visualization of the worker agent based on TensorBoard.

In order to verify the feasibility of the EABRA algorithm, we compare it with the Random (mobile terminals perform actions randomly). In Fig. 4, we compare the loss functions of the two algorithms, and we can see that the EABRA algorithm converges faster, and the performance of the Random is poor, which can hardly converge.

Figure 5 shows the impact of the computing capability of edge node on the total energy consumption. It can be seen that for all schemes, the total energy consumption increases with the increase of f_n. This is because as the CPU frequency of the edge node increases, although the calculation rate will increase, the communication overhead will also increase, which will lead to additional energy consumption, thus increasing the total energy consumption of the system. Figure 6 shows the effect of transmission power P

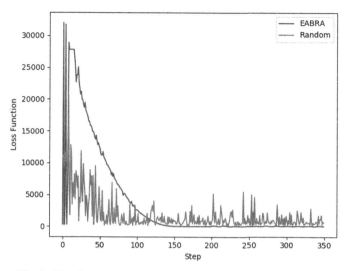

Fig. 4. The change curve of loss function based on A3C algorithm.

on the average reward. It can be seen that the performance of EABRA is always the best, because we jointly optimize the task offloading decision and the transmission power allocation decision, and find the optimal resource allocation scheme base on the DRL algorithm.

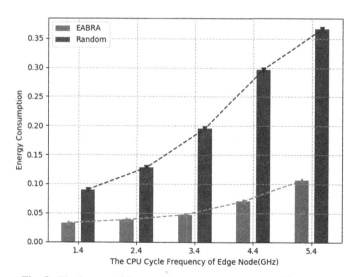

Fig. 5. The impact of f_n on total energy consumption of the system.

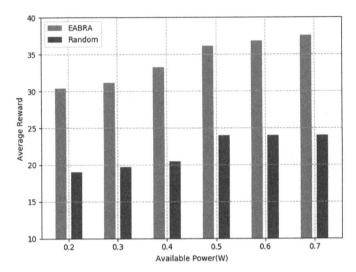

Fig. 6. The impact of P on average reward of the system.

5 Conclusion

This paper proposes a blockchain-based edge-terminal collaborative resource allocation architecture, including MEC system and blockchain system. Local mobile terminal computing mode and edge node collaborative computing mode are used to process the mining task of blockchain system, so as to ensure the trust of terminal authentication information. In order to optimize the total energy consumption of the system, we model the problem as an MDP problem, jointly optimize the offloading decision and power allocation decision, and propose the EABRA algorithm with deep reinforcement learning for solution. The simulation results clearly show the superiority of EABRA algorithm. Under different parameter settings, the EABRA algorithm has faster convergence speed and better performance.

Acknowledgment. This work is supported by the Science and Technology Project of State Grid Corporation of China: Research on Key Technologies of dynamic identity security authentication and risk control in power business (SGHEXT00YJJS1900050).

References

1. Roman, R., Zhou, J., Lopez, J.: On the features and challenges of security and privacy in distributed internet of things. Comput. Netw. **57**(10), 2266–2279 (2013)
2. Rashid, M.A., Pajooh, H.H.: A security framework for iot authentication and authorization based on blockchain technology. In: 2019 18th IEEE International Conference on Trust, Security and Privacy in Computing and Communications/13th IEEE International Conference on Big Data Science and Engineering (TrustCom/BigDataSE), Rotorua, New Zealand, pp. 264–271 (2019)

3. Guo, S., Hu, X., Guo, S., Qiu, X., Qi, F.: Blockchain meets edge computing: a distributed and trusted authentication system. IEEE Trans. Ind. Inform. **16**(3), 1972–1983 (2020)
4. Neisse, R., Hernández-Ramos, J.L., Matheu, S.N., Baldini, G., Skarmeta, A.: Toward a blockchain-based platform to manage cybersecurity certification of IoT devices. In: 2019 IEEE Conference on Standards for Communications and Networking (CSCN), Granada, Spain, pp. 1–6 (2019)
5. Cui, Z., et al.: A hybrid blockchain-based identity authentication scheme for Multi-WSN. IEEE Trans. Serv. Comput. **13**(2), 241–251 (2020)
6. Thakker, J., Chang, I., Park, Y.: Secure data management in internet-of-things based on blockchain. In: 2020 IEEE International Conference on Consumer Electronics (ICCE), Las Vegas, NV, USA, pp. 1–5 (2020)
7. Liu, M., Yu, F.R., Teng, Y., Leung, V.C.M., Song, M.: Computation offloading and content caching in wireless blockchain networks with mobile edge computing. IEEE Trans. Veh. Technol. **67**(11), 11008–11021 (2018)
8. Feng, J., Yu, F.R., Pei, Q., Chu, X., Du, J., Zhu, L.: Cooperative computation offloading and resource allocation for blockchain-enabled mobile-edge computing: a deep reinforcement learning approach. IEEE Internet Things J. **7**(7), 6214–6228 (2020)
9. Mnih, V., et al.: Asynchronous methods for deep reinforcement learning. In: International Conference on Machine Learning, pp. 1928–1937 (2016)

Internet of Power Things and Big Data

IPv6 Header Compression Scheme for Power Internet of Things

Wang Xiaoyu[1](✉), Lu Xu[1], Liu Chuan[2,3], Tao Jing[2,3], and Liang Zhonghua[1]

[1] China Academy of Information and Communications Technology, Beijing 100191, China
[2] Global Energy Interconnection Research Institute Co., Ltd., Nanjing 210003, China
[3] State Grid Laboratory of Electric Power Communication Network Technology, Nanjing 210003, China

Abstract. In power Internet of Things, to enable the underlying nodes to communicate with an IPv6 network, a 6LoWPAN adaptation layer must be supported through a gateway to seamlessly connect them to implement IPv6 and IEEE802.15.4 protocols. The 6LoWPAN adaptation layer can improve the transmission efficiency of data packets by compressing IPv6 data headers. According to the characteristics of energy metering systems in smart grids, UDP is used as a default transmission layer protocol. Based on existing header compression schemes LoWPAN_HC1 and LoWAPN_IPHC, an adaptive hybrid header compression scheme LoWPAN_HC_Energy was proposed for energy metering systems. The experimental results illustrated that the scheme performs satisfactorily in both local and global links, and its compression efficiency is approximately 2% higher than that of existing compression schemes .

Keywords: Energy metering system · Gateway · 6LoWPAN · Header compression

1 Introduction

With the rapid development of computer and embedded technology, the Internet of things (IoT) is becoming increasingly popular in people's lives [1] and the concepts of smart earth and city are gradually emerging [2]. In smart grids, an energy metering system is a subset of smart energy, which is used to solve energy metering and metering data management functions in smart energy [3].

The main component of energy metering systems is an IoT system, which is composed of three parts: a bottom sensor, gateway, and user [4]. The sensor in the system is mainly responsible for collecting energy measurement data, and the gateway is responsible for collecting and processing underlying data and displaying it to the user. The gateway plays a bridging role in the system, which is responsible for the connection between an internal wireless sensor network and external Internet users [5].

Fund project: State Grid Corporation Headquarters Technology Project (5700-201999496A-0-0-00).

With the emergence of an IoT wave, under the demand of 'Internet of everything', the IPv4 protocol can no longer afford the large number of network address requirements and the IPv6 protocol has been ushered in an excellent opportunity of development and application. To make metering systems compatible with future network protocols, gateways should provide system support for IPv6 to seamlessly connect underlying sensor nodes with IPv6 networks. Due to an incompatibility between IPv6 and IEEE802.15.4 protocols, IPv6 adaptation over low power wireless personal area networks (6LoWPAN) layer [6] must be completed through the gateway. Figure 1 presents the location of the 6LoWAPN adaptation layer in the network protocol layer and efficient conversion between IPv6 packets and IEEE802.15.4 frames.

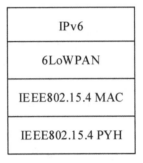

Fig. 1. Network protocol layer

Figure 2 illustrates the IPv6 packet header. The maximum transmission unit (MTU) length is 1280 byte, and the IEEE802.15.4 MTU is 127 byte, which does not meet IPv6 requirements. Therefore, when the packet transmitted through IPv6 is greater than its MTU, it must be fragmented and reorganised using 6LoWPAN.

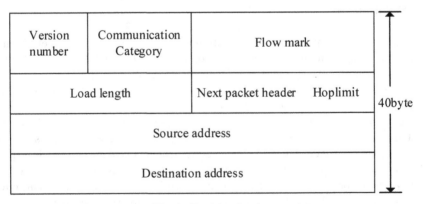

Fig. 2. IPv6 data header

In addition, after the removal of necessary fields, the maximum payload in IEEE802.15.4 and fixed header length in IPv6 are 93 and 40 bytes, respectively. If it is directly transmitted without compression, the transmission efficiency obtained is low. Therefore, header compression is required to improve the data transmission efficiency.

Studies on 6LoWPAN mainly include implementing 6LoWPAN and improving the 6LoWPAN efficiency in packet fragmentation and reassembly, header compression, and other aspects. Wang Xiaoan proposed a design and implementation of 6LoWPAN sensor nodes [7] and applied it to the real-time monitoring system of the agricultural environment [8]. Geng Daoqu studied the packet scheduling method of 6LoWPAN access to IPv4 [9] and proposed an adaptive joint gateway that can simultaneously communicate with both IPv4 and IPv6 networks [10]. Liang Bao proposed a TCP-based 6LoWPAN header compression scheme [11], and then He Donghui proposed a universal header compression scheme: 6LoWPAN_GHC [12]. In the case of bandwidth and limited resources, effective packet header compression of IPv6 is very important. [13] Proposed an improved header compression scheme, which can better compress the Pv6 multicast address, ICMP header and routing extension header [14]. In an open stream network with satellite links, a method with a high header compression ratio is proposed. MAC headers can also be compressed in OpenFlow networks with satellite links [15, 16]. In order to improve the efficiency of data transmission, IPv6 data header compression is necessary.

This paper combines the characteristics of the energy metering system to modify the available header compression schemes LoWPAN_HC1 and LoWAPN_IPHC and presents a design of an adaptive header compression scheme LoWPAN_HC_Energy for energy measurement systems.

2   Introduction and Modification of Header Compression Scheme

2.1   Introduction to LoWPAN_HC1

In 2007, the 6LoWAPN working group proposed LoWPAN_HC1 [13] that is a header compression scheme suitable for local link networks. This scheme sets the specific compression format of the data packet to the 1-byte LoWAPN_HC1 compression control header field. In optimal compression, the fixed header information of 40 bytes can be compressed to 2 bytes. Figures 3 and 4 present the overall coding format of LoWPA_HC1 and specific fields of 'HC1 coding', respectively.

01000010	HC1 encoding	HC2 encoding	Hop limit

Fig. 3. LoWPAN_HC1 compression scheme

In HC1 encoding, an 'IPv6 source/destination address field is used to determine an address compression mode in the gateway header. When set to 11, the IP address can be calculated from a network prefix and MAC layer address. The network address is composed of the network prefix and IID compression and is directly omitted in the

IPv6 source address	IPv6 destination address	Transmission type and stream label	Next header type	HC2 encoding type

Fig. 4. HC1 encoding field

IPv6 header. A 'transmission type and flow label' field determines whether transmission type and flow label field in the IPv6 data header are compressed. A value of 0 indicates no compression, and a value of 1 denotes the default route, which compresses it. The 'next header type' field indicates the next header type in a data packet. 00, 01, 10, and 11 indicate no compression, NH is the IDP header, NH is the ICMP header, and NH is the TCP header, respectively. An 'HC2 encoding type' field indicates whether NH is compressed. When HC2 encoding field is 0, no compression is performed. The HC2 encoding field of 1 indicates that UDP is compressed. The HC1 encoding field of '11111011' indicates that the compression target is a local data packet, and the transport layer uses a UDP protocol and compresses it. At this time, LoWPAN_HC1 exhibits the highest compression efficiency. The HC2 encoding field is used to set the compression parameters of UDP. Figure 5 illustrates the encoding field.

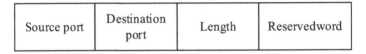

Source port	Destination port	Length	Reservedword

Fig. 5. HC2 encoding field

2.2 Header Compression Scheme for Energy Metering System LoWPAN_HC1_Energy

In the HC1 encoding field of LoWAPN_HC1, the type of the adjacent joint part is set through the 'next header type' field, and UDP compression is set through the HC2 encoding type. In energy measurement systems, we recommend to use UDP as default transport layer protocol and to compress it. Figure 6 presents the encoding format used by the protocol.

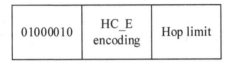

01000010	HC_E encoding	Hop limit

Fig. 6. LoWPAN_HC1_Energy compression scheme

Figure 7 presents the encoding format of the HC_E field. The first 5 bits are similar to the HC_1 encoding field, which is used to set the IPv6 source/destination address, transmission type, and stream label compression. Because UDP is used as the default transport layer protocol, the setting of the last three digits of LoWPAN_HC1 for the protocol type can be replaced by compression setting for the UDP header. A hop limit remains constant.

In this manner, the UDP protocol can be compressed by default, and the compressed header length can be reduced by 8 bits compared with its reduction using the header compression scheme LoWAPN_HC1.

IPv6 source address	IPv6 destination address	Transmission type and stream label	UDP source port	UDP destination port	UDP length

Fig. 7. HC1_E coding field

2.3 Introduction to LoWPAN_IPHC

LoWPAN_HC1 only exhibits a good compression effect on the local link. In the global network, because source and destination addresses use different routing prefixes, LoW-PAN_HC1 cannot effectively compress the IPv6 address, which results in a highly reduced compression efficiency of the scheme.

The 6LoWAPN working group proposed a new header compression scheme LoWAPN_IPHC in 2011 [17].

The LoWPAN_IPHC adaptation layer header is the global shared Context Table. Some frequently-occurring IPv6 address prefixes can be saved, and the address in the packet header is represented by the index of the table. Context Identifier Extension (CID) replaces the public address prefix, thereby reducing data packet size. Although the LoWPAN_IPHC adaptive layer header can increase the application layer space, Context Table can only store 16 address prefixes [18].

As a new header compression scheme, LoWPAN_IPHC overcomes shortcomings such as LoWPAN_HC1 can effectively compress only local network data packets and can more effectively compress various communication scenarios. LoWPAN_IPHC sets the parameters of IPv6 header compression through the LoWPAN_IPHC field, and Fig. 8 presents the specific contents of the field.

1	1	1	Communication type and flow label	Next head	Maximum hop limit
Context	Source address	Source address compression mode	Whether to multicast	Destination address	Destination address compression mode

Fig. 8. LoWPAN_IPHC coding field

Different from LoWAPN_HC1, the scheme compresses the maximum hop count HLIM field to 2 bits, where 00 indicates that the maximum hop limit is not compressed and the 8-bit hop limit in the IP header is retained and 01/10/11 indicates that the hop limit field is compressed and set to the common 1/64/255 hop, respectively. In addition, the scheme uses the source address compression control (SAC)/source address compression mode (SAM) to set the compression method of the source address, and either multicast, destination address compression, or destination address compression mode to set the compression method of the target address. The SAM set to 0 indicates that the stateless compression method is used. At this time, the compression method used in SAC is similar to LoWPAN_HC1. The SAM of 1 indicates the use of a context-based compression method for compression. Table 1 presents the specific meaning of each field in SAC at this time.

Table 1. SAM field meaning

SAM	Field meaning
00	Undefined
01	Compression, using a 64-bit address, specific address information can be obtained from upper and lower layers
10	Compression, using a 16-bit address, specific address information can be obtained from upper and lower layers
11	Compression, address omitted, specific address information can be obtained from upper and lower layers

The method of destination address compression is similar to that of source address compression and is not repeated here. For details, please refer to the document RFC6282 [19].

2.4 Header Compression Scheme for Energy Metering System LoWPAN_IPHC_Energy

We modified the compression scheme LoWPAN_IPHC according to the characteristics of the default UDP protocol adopted by the energy measurement system, and the header compression scheme (Fig. 9) can be achieved. Figure 10 presents the specific coding field of LoWPAN_IPHC_Energy.

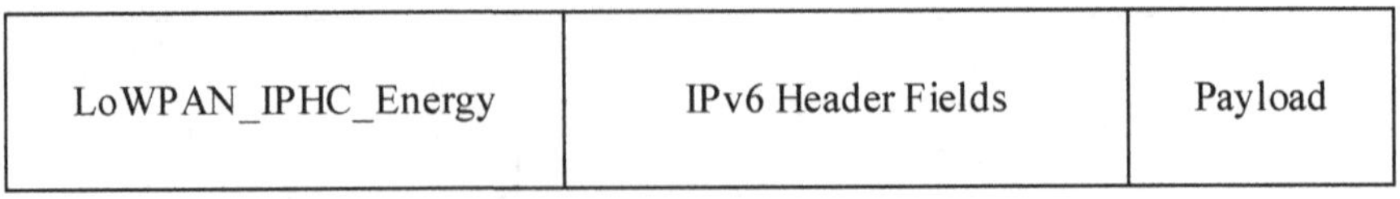

LoWPAN_IPHC_Energy	IPv6 Header Fields	Payload

Fig. 9. LoWPAN_IPHC_Energy compression scheme

1	1	1	Communication type and flow label	Port compression mode P	Maximum hop limit
Context	Source address	Source address compression mode	Whether to multicast	Destination address compression	Destination address compression mode

Fig. 10. LoWAPN_IPHC_Energy coding field

When UDP is compressed in LoWPAN_IPHC, the specific parameters of header compression are set through the LoWPAN_NHC field, which mainly includes the setting of two parameters: 'checksum' and 'port compression mode'. In the energy metering system compression scheme LoWPAN_IPHC_Energy, the 'checksum' field setting is deleted. By default, checksum is used to check the system and HLIM is further compressed. The setting of checksum field to 0 indicates no compression; otherwise it will jump the maximum. The number limit is set to 64. In this manner, the compressed header length is reduced by 8 bits in LoWPAN_IPHC_Energy compared with in LoWAPN_IPHC.

3 Adaptive Header Compression Scheme

Header compression schemes LoWPAN_HC1_Energy and LOWPAN_IPHC_Energy are slightly modified versions of LoWAPN_HC1 and LoWPAN_IPHC created to slightly improve the compression effect. During communication in local networks, the compression effect of the LoWAPN_HC1 scheme is better. During communication on global networks, LoWPAN_IPHC can considerably effectively compress IPv6 headers. Therefore, two compression methods can be integrated to compress data packets. In the energy measurement system, if a current communication range belongs to local communication, LoWPAN_HC1_Energy is used for header compression, otherwise LoWAPN_IPHC_Energy is used for header compression. Adaptive compression scheme LoWPAN_HC_Energy can further improve the compression efficiency by integrating two compression schemes.

1	1	1	Type control C	LOWPAN_X_Engery

Fig. 11. LoWPAN_HC_Energy message format

Figure 11 presents the data message format of LoWPAN_HC_Energy. The first three digits '111' are used as the category identifier of the data packet, and the fourth digit 'category control' is used to select the type of the compression scheme to be used. When set to 0, the LoWPAN_HC1_Energy compression scheme is used, otherwise the LoWAPN_IPHC_Energy compression scheme is used. Next 12 bits are used to set the specific compression method of the scheme. Figures 12 and 13 present the meaning of specific fields.

When the 'Type Control' field is 0, LoWPAN_HC1_Energy is used for compression. The compression scheme uses the same coding format as the HC_E field does in LoWPAN_HC1_Energy, and 4 bits are added beforehand as reserved bits to obtain the 12-bit LoWAPN_X_Energy field to set compression parameters.

Reserved bit	IPv6 source address	IPv6 destination address	Communication type and flow label	UDP source port	UDP destination port	UDP length

Fig. 12. LoWPAN_HC_Energy compression format 1

When the 'type control' field is 1, LoWPAN_IPHC_Energy is used for compression, the HLIM field is omitted, and the maximum hop limit field is not compressed. Other fields remain reserved, and the 12-bit LoWAPN_X_Energy field completes the control of the compression parameters.

1	1	1	Communication type and flow label	Port compression mode	Maximum hop limit
Context	Source address	Source address compression mode	Whether to multicast	Destination address	Destination address compression mode

Fig. 13. LoWPAN_HC_Energy compression format 2

4   Experiments

For comparison, a compression ratio was used as a compression scheme index. Let the size of the data packet before and after compression be I and C, respectively, and the compression ratio P can be calculated using Eq. 1.

$$P = 1 - C/I \tag{1}$$

The compression efficiency of compression scheme LoWAPN_HC1_Energy was related to whether the data packet is a local link. The 'probability of transmitting a data packet as a local link' was considered as an abscissa, and the 'compression rate' was considered an ordinate. Under various compression schemes with different probabilities, the observed data packet was the compression effect of the local link. Table 2 present the results and the specific data results.

Table 2. Compression effect of each header compression scheme

Compression scheme	Average compression ratio	Maximum compression rate	Minimum compression rate
LoWPAN_HC1	12.98	17.42	9.39
LoWPAN_HCI_Energy	14.88	19.32	11.29
LoWPAN_IPHC	15.88	16.64	14.91
LoWPAN_IPHC_Energy	17.76	18.56	16.77
LoWPAN_HC_Energy	17.96	18.92	17.13

Compared with LoWPAN_HC1, the compression rate of LoWPAN_HC1_Energy increased by approximately 1.90%; compared with LoWPAN_IPHC, the compression rate of LoWPAN_IPHC_Energy increased by approximately 1.88% (Table 2).

With an increase in the probability of the transmission packet of being a local link, the compression effect of LoWAPN_HC1 and LoWPAN_HC1_Energy gradually increased. Simultaneously, the compression effects of LoWPAN_IPHC and LoWPAN_IPHC_Energy were not affected. The compression effect of the adaptive compression scheme was always relatively ideal and was always large for LoWPAN_HC1_Energy and LoWPAN_IPHC_Energy.

5 Conclusions

In the context of the continuous development of smart cities and smart energy technologies, the IPv6 protocol is required by the IoT system for network scalability. Because the IPv6 data header is considerably large, completing the efficient transmission of IPv6 is difficult through an IEEE802.15.4 payload. Therefore, to improve the data transmission efficiency, IPv6 data header compression is necessary.

In the energy metering system, UDP is used as the default transport layer protocol and performs compression, which can improve the compression efficiency of existing compression schemes. LoWPAN_HC_Energy combines the characteristics of existing two header compression schemes, the adaptive method can be adopted to compress different types of data headers differently, and the compression effect of the data packet can further be improved.

References

1. Sestino, A., Prete, M.I., Piper, L., Guido, G.: Internet of Things and Big Data as enablers for business digitalization strategies. Technovation 102173 (2020)
2. Shafique, K., Khawaja, B.A., Sabir, F., et al.: Internet of Things (IoT) for next-generation smart systems: a review of current challenges, future trends and prospects for emerging 5G-IoT scenarios. IEEE Access **1**(8), 23022–23040 (2020)
3. Jia, H., Gai, Y., Xu, D., et al.: Link importance-based network recovery for large-scale failures in smart grids. Wirel. Netw. (2020)

4. Sun, Q., Li, H., Ma, Z., et al.: A comprehensive review of smart energy meters in intelligent energy networks. IEEE Internet Things J. **3**(4), 464–479 (2016)
5. Gokbayrak, K.: Robust gateway placement in wireless mesh networks. Comput. Oper. Res. (97), 84–95 (2018)
6. Tsiatsis, V., Karnouskos, S., Höller, J., et al.: Internet of Things (Second Edition) [B] Technology Fundamentals, pp. 67–126. Academic Press, America (2019)
7. Wang, X., Gao, D.: Design and implementation of 6LoWPAN sensor node. J. Sensor Technol. (10), 1501–1504 (2010)
8. Wang, X., Yin, X.: Real-time monitoring system for agricultural environment based on 6LoWPAN wireless sensor network. Trans. Chin. Soc. Agric. Eng. (10), 224–228 (2010)
9. Geng, D., Dai, F., Li, X., et al.: Research and implementation of packet scheduling in 6LoWPAN access (IPv4) Internet. J. Sens. Technol. (12), 1752–1756 (2013)
10. Geng, D., Chen, H., Chai, J., et al.: Design and implementation of an adaptive unified gateway for 6LoWPAN to access the Internet. J. Sensor Technol. (3), 416–423 (2015)
11. Liang, B., Li, H., Jia, J., et al.: A TCP header compression scheme based on 6LoWPAN. Appl. Comput. Syst. (6), 119–122 (2013)
12. He, D., He, T., Dai, J.: Study on 6LoWPAN_GHC header compression algorithm. Ind. Control Comput. (6), 110–112 (2015)
13. Huiqin, W., Yongqiang, D.: An improved header compression scheme for 6LoWPAN networks. In: 2010 Ninth International Conference on Grid and Cloud Computing, Nanjing, pp. 350–355 (2010)
14. Venmani, D.P., Duprez, M., Ibrahim, H., et al.: Impacts of IPv6 on robust header compression in LTE mobile networks (2012)
15. Niu, Y., Wu, C., Wei, L., Liu, B., Cai, J.: Backfill: an efficient header compression scheme for OpenFlow network with satellite links. In: 2016 International Conference on Networking and Network Applications (NaNA), Hakodate, pp. 202–205 (2016)
16. Sun, S., Chen, Y., Piao, Z., Zhang, J.: Vessel AIS trajectory online compression based on scan-pick-move algorithm added sliding window. IEEE Access **8**, 109350–109359 (2020)
17. Montenegro, G., Kushalnagar, N., Hui, J.: Transmission of IPv6 packets over IEEE802.15.4 networks. IETF RFC 4944 (2007)
18. Cui, L., Hua, G., Lu, N.: A Dynamic 6LoWPAN Context Table Maintaining algorithm. In: 2013 9th International Wireless Communications and Mobile Computing Conference (IWCMC), Sardinia, pp. 1458–1463 (2013)
19. Hui, J., Thubert, P.: Compression format for IPv6 datagrams over IEEE 802.15.4 based networks. IETF RFC 6282 (2011)

IPV6 Address Configuration Method in 6LoWPAN Oriented to the Internet of Power Things

Lu Xu$^{(\boxtimes)}$, Li Jianwei, Jiang Hao, Luo Dan, and Cao Han

China Academy of Information and Communications Technology, Beijing 100191, China

Abstract. Taking the location information as the focus, we analyze the accurate algorithm of the location-based 6LoWPAN network to automatically configure the address, and conducts a feasibility study. The main content includes the establishment of a LINA network architecture based on location information. Based on the location information of the node, the router at the edge is used as the center, and the plane projected on the 6LoWPAN network is used as the medium to form many grids and achieve the goal of rapid clustering; each grid is independent of each other and can configure the address effectively at the same time, so that the delay of the network can be greatly reduced; the feasibility of the address configuration scheme is verified by comparing the overhead of the traditional address configuration scheme .

Keyword: Power Internet of Things · 6LoWPAN · IPV6 · Address configuration

1 Introduction

The power Internet of Things has developed rapidly, and the current IPv4 protocol cannot satisfy the increasingly diversified needs; therefore, the new IPv6 protocol came into being. It has a 128-bit address and rich resources. It can basically satisfy the requirements of uniform addressing of nodes for the power Internet of Things on a theoretical level [1]. However, the current Internet of Things has the characteristics of low broadband and power consumption, which hinders the effective use of the IPv6 protocol. IETF proposed the 6LoWPAN protocol in response to the above situation, which means that devices with low-level power consumption provide strong support for the realized IP functions, that is, the low-power network of the power Internet of Things and the Internet based on IPv6 are seamlessly connected, and unified Addressing function. 6LoWPAN has become a crucial key technology in the development of the field of power Internet of Things, making the power Internet of Things more widely used [2].

In recent years, the main content of the industry research is the 6LoWPAN low-power configuration solution for the underlying network address. According to the need for

Fund project: State Grid Corporation Headquarters Technology Project (5700-201999496A-0-0-00).

M. Cheng et al. (Eds.): SmartGIFT 2020, LNICST 373, pp. 117–127, 2021.
https://doi.org/10.1007/978-3-030-73562-3_10

DAD conflict detection, it is divided into the following types (1) Conflicting configuration address protocol. It repeats detection for the generated address and makes its unique attributes verified; (2) Configure the address protocol without conflict. It uses neighbor proxy methods to assign and decentralize address tasks to network nodes.

Based on the allocation of IP addresses to nodes, the low-power bottom nodes in the 6LoWPAN network and the corresponding devices in the IP network can communicate with each other [3]. Based on a conventional wired network, the address is effectively configured through IPv4 and IPv6 protocols; the configuration method covers the following types [4]: (1) Automatic configuration, whether there is a central server or not, it can be divided into the two types of stateful and stateless; (2) Manual configuration. Both IPv4 and IPv6 support it. When the number of network nodes is relatively small, the address is manually configured to each node. If the number of nodes is large and the IPv6 address has extremely long, manual configuration mode is not suitable, so it will be a crucial link to automatically obtain IP when loading nodes in the network. In the past, when the IP network realized the automatic configuration address function, IPv4 provided a protocol for stateful automatic configuration address, which was realized by the DHCP server. After further improvement and improvement, IPv6 as a new IP protocol that [5] it provides support for stateless addresses while making the configuration extremely flexible and convenient. The automation of stateful address configuration is usually based on the DHCP or DHCPv6 server to configure the IP address of the terminal device to the Internet, which means that the server/client mode of operation is adopted [6]. The server is the center to store the status information of each node and the corresponding addresses that have been allocated to it, and the allocation table is constantly updated and improved. If a new node is added to the network, a request for address application will be made to the central server. After receiving the corresponding request, the server will assign the address to the corresponding node [7]. After obtaining the address from the server, the node returns to the server and clarifies the information. After the server receives the corresponding information, it records the corresponding information of the assigned node in the address allocation table [8]. In smart grids, an energy metering system is a subset of smart energy, which is used to solve energy metering and metering data management functions in smart energy [9].

2   Introduction of 6LoWPAN

2.1   IPv6 Addressing of 6LoWPAN Network

The Internet of Power Things has developed rapidly, and the currently adopted IPv4 protocol cannot satisfy the increasingly diversified needs; therefore, the new IPv6 protocol came into being, with a 128-bit address and abundant resources, which can basically provide every drop of water a unique identifier based on the theory [5]. It is in line with the requirement that IoT nodes can be addressed uniformly. However, the current embedded network has the characteristics of low broadband and power consumption, which hinders the effective use of the IPv6 protocol. The IETF proposed the 6LoWPAN protocol in response to the above situation, which means that devices with low-level power consumption provide strong support for the realized IP functions, that is, the low-power network of the Internet of Things and the Internet based on IPv6 are seamlessly connected, and unified search is realized. The function of 6LoWPAN has become a crucial key technology in the development of the Internet of Things and makes the Internet of Things more widely used [7]. For the IP-based Internet, in order to communicate with each other normally and smoothly, the legality of IP addresses needs to be unique [8]. As a new model for constructing networks, 6LoWPAN needs to effectively solve the problem of configuring IP addresses, so that it can be used freely and connected with other networks. Based on the existing IP network, nodes generally obtain corresponding addresses through existing state or stateless configuration addresses. If the solution of stateful address configuration is adopted, it refers to the effective configuration and management of corresponding addresses for network terminals via the mode of DHCPv6 and Internet broadcasting. IPv6 address covers the link between prefix and interface identification. The length of the address is 128 bits, and each segment is 16 bits. Each segment can be turned into hexadecimal numbers and separated; IPv6 addresses can be divided into the following types: first, unicast. The destination address is effectively specified for a host or router, that is, the identifier corresponds to a single interface; second, anycast. The destination station shares a set of nodes in a certain address, selects the shortest path through datagrams, and then delivers to the group members; third, multicast. The destination station is a group of addresses, and the datagram is delivered to the group members in the form of multicast or broadcast. In smart grids, an energy metering system is a subset of smart energy, which is used to solve energy metering and metering data management functions in smart energy [9].

IP addressing combines the prefix of the underlying wireless personal area network with the link layer address of the interface to generate an IPv6 address. The length of the address is 128 bits, which are the routing prefix and the interface identifier. The nodes of the same 6LoWPAN network share the prefix of the route, which can be obtained by the broadcast message of the link; the 16-bit short address is generated according to the 64-bit interface identifier. The following figure describes the IPv6 structure diagram (Fig. 1).

Global routing prefix	0000	00FFFE00	16-bit short address
64bit	16bit	32bit	16bit

Fig. 1. IPv6 structure diagram

3 Location-Based Information Network Architecture

In the power Internet of Things, the location information between nodes can express the distance and directionality of nodes in the physical space, and the nodes themselves have natural similarities and differences. Through in-depth analysis of the location information, some common problems can be solved more effectively. For example, through the similarities and differences of the location information, the corresponding address identification can be formed through the information, and the task of stateless automatic address configuration can be conveniently completed. In addition, through the distance and directionality of the location information in the node address information, the route can be easily determined, so that the overhead of routing addressing is greatly reduced. Therefore, in the process of 6LoWPAN network application, the design idea of addressing through location information has remarkably huge promotion and application prospects and needs further development.

This article applies the distance, direction and similarities and differences of node location information, and on the basis of 6LoWPAN network characteristics and technical research, we analyze the address configuration scheme based on location information. Figure 2 is a network topology diagram based on location information:

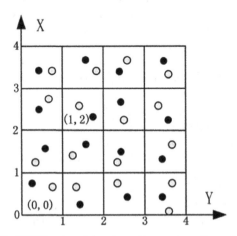

Fig. 2. 6LoWPAN network topology diagram based on network grid

When constructing the relevant network framework for location information, the 6LoWPAN network must first be divided into a number of virtual network two-dimensional grids according to the number of nodes and network coverage [37]. The area of one grid is represented by S, following with the formula $S = (mr)^2$. r always represents the wireless transmission radius of the node, and m is customized according to the number of nodes in the grid (it is a positive integer). We set the coordinates of the edge router (0, 0) and each grid is identified by the coordinates (X, Y). The node is connected to the network. A specific positioning method is firstly used to obtain its own positioning coordinates (Lx, Ly), and the following formula is performed to calculate the corresponding deleted grid coordinates. See the following formula for details:

$$X = \left[\frac{L_X}{\sqrt{S}} \right] \tag{1}$$

$$Y = \left[\frac{L_Y}{\sqrt{S}} \right] \tag{2}$$

The solution in this paper divides the short address (16 bits) of the 6LoWPAN network, that is, the node ID into three parts: internal ID, ordinate, and abscissa. The address length of the abscissa is i bit, and the address length of the ordinate is j bit.

We assume i and j are positive integers, and their values depend on the density and number of nodes distributed in the network. Figure 3 shows the IPv6 address structure is designed by the above network architecture. It is a network architecture based on location information. Its purpose is to complete the function of converting location information to address space, so that the location information of the node is included in the node's corresponding IPv6 address, which provides great convenience for the specific application of the 6LoWPAN network based on location information. In addition, according to the natural similarities and differences of node location information, the network achieves the function of rapid clustering without other address configuration and complicated clustering algorithms. Moreover, under the condition of a large-scale network and a large number of nodes, a very large network can be divided into multiple subnets, and all subnets are independent of each other, and address configuration can be completed simultaneously, which greatly reduces network delay and expenses.

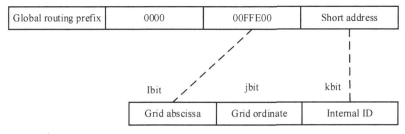

Fig. 3. IPv6 address structure based on location information address automatic configuration scheme

3.1 Node Type

In this solution, the relevant nodes in the access network specifically include three categories, and the various states of the nodes correspond to the various categories of the nodes. In the case of different states, the functions undertaken by the nodes also have certain differences. This program mainly involves the following four types of nodes:

(1) Newborn node: The node has not yet obtained a reasonable IPv6 address. It has just started to access the network, and obtain the corresponding grid ID according to its own location information, and determine the grid where it is located.

(2) Normal node: It has no redundant address, but is configured with a legal IPv6 address corresponding to it, and it cannot be configured with an IPv6 address to a new node.

(3) Proxy node: It has redundant addresses and is configured with a legal IPv6 address corresponding to it. The address space is greater than 1, and it can be configured with IPv6 addresses to new nodes.

(4) Head node: All deleted grids have 1 head node, and 1 grid is equivalent to 1 subnet. This head node is equivalent to an IPv6 router. Its function is to configure the corresponding address to the internal node of the deleted grid to communicate with each other.

When the new node approaches the network, it can obtain the corresponding grid ID. First, it will send out an address application to neighboring nodes within one hop of the surrounding area, and the proxy node will respond after receiving the request. If the neighboring nodes in the surrounding one-hop interval are all new nodes or normal nodes (that is, there is no free address), the scope will be expanded and the address application will continue. If the proxy node allocates the corresponding address to the new node with only one redundant address, then after the redundant address is allocated to the new node, the node becomes a normal node because it no longer has a spare address. The proxy node itself has redundant addresses. After configuring all the redundant addresses in the process of configuring other addresses, it will become a normal node. In the same way, when a normal node obtains other spare addresses in the network, it can also become a proxy node. If normal nodes and proxy nodes are separated from the network for various reasons, they become invalid nodes. The schematic diagram of node status changes is shown in Fig. 4.

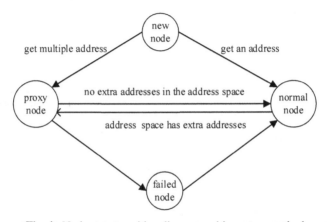

Fig. 4. Node state transition diagram on binary tree method

This configuration address scheme divides the 6LoWPAN network into several network grids (virtual) according to the location information of the nodes, and realizes the function of rapid clustering. It is assumed that all deleted grids have their own corresponding cluster head nodes (that is head node). The network architecture is the same as shown in Fig. 3. After the head node is connected to the network, it will first obtain the location information and grid ID corresponding to itself, confirm the grid where it is located, and set the internal ID corresponding to itself to 1. It combines the two stages to form a short address (16 bits) and then configure an IPv6 address, which is the same as shown in Fig. 2.4. At this time, the network initialization is complete. The head node uses the random dichotomy to configure the nodes in the grid to internal IDs, and when setting the parameters of the random dichotomy, a certain address space is set according to the number of nodes near the node that have not been configured with IP addresses to reduce the probability of unsuccessful configuration in the range of high node density in the interval during the process of address configuration, and an effective address recovery method is also applied. The node adopts the above method to automatically obtain the

16-bit short address that is allowed to be used inside the 6LoWPAN network, and does not need to detect whether the address conflicts. In addition, all grids are independent of each other, and several grids can be configured with addresses at the same time, which reduces the overhead and the delay in the process of addresses configuration.

4 Address Allocation Method

4.1 IP Address Allocation

After a new node joins the network, it needs to extract the horizontal and vertical coordinates of the deleted grid according to its own positioning information, and set a head node in each area. We set the ID of the head node to be 1. The head node covers the entire address space. When configuring the address, the configuration should be started from the head node. In the following links, a new network node is added to obtain the internal ID process. In the early stage when it does not get a unique address, the node uses its own MAC address as a temporary address when communicating.

(1) After the new node A obtains the grid coordinates, it will send the address request addr_request control packet to the neighbor nodes within one hop range, and set the hop count R (positive integer) and the number of transmissions to 1, and configure the corresponding Offer_timewait waiting clock.

(2) After the neighbor node within one hop receives the addr_request control packet, if it is judged that the received data packet is a normal node, it will discard it; if it is judged that the received data packet is a new node, it will send the addr_reply1 data packet. If it is determined that the received data packet is a proxy node, the data packet addr_reply2 is sent.

(3) After Offer_timewait waits for the time set by the clock, the addr_reply1 and addr_reply2 data packets replies from neighbor nodes are checked by node A, and the number N of new neighbor nodes within one hop is calculated, and the replying proxy node is compared with the number of free addresses W. It sends the selc_reply data packet to the proxy node B with the largest number of free addresses. If the data packet addr_reply2 from the proxy node is not received, it is confirmed that there are no neighbor nodes within one hop. Address (6).

(4) The selc_reply data packet returned by the new node A is received by the proxy node B with the largest number of free addresses. At the same time, the value of the new number field in the packet is checked, that is, the new node N with no address configured in the range of the new node A is checked. According to the value of N and the number of free addresses W (W is a positive integer ≥ 1), node B can generate a random number M to determine the number of addresses allocated to new node A.

(5) After the addr_ack control data packet sent by the proxy node B is received by the new node A, it can obtain its own internal ID, construct the 16-bit short address together with the internal ID and the grid ID, and then calculate with the obtained corresponding IPv6 address. At the same time, node A checks the number M of addresses assigned to it by proxy node B. If it is $M \geq 2$, it will mark itself as a proxy node, and if $M = 1$, it will mark itself as a normal node.

(6) When node A cannot find an available address in the neighboring node within one hop, it will send the addr_reques address to expand the range of the control packet, and also actively increase one hop, and at the same time, the number of transmissions and the number of channels R are both set to 2.

(7) After the neighbor node H receives the addr_reques packet, when it is a normal node or a new node, it will discard it without processing. When it is a proxy node, it will reply the addr_Reply2 packet to the node. It set the hop count R value in the data packet to 2, and the free address number fre-addr number field value to 0.

(8) After node A transmits the selc_reply data packet to the proxy node H that answers the fastest, it also sets the value of the new number field of the new node to 0.

(9) After receiving the selc_reply data packet, the proxy node H allocates its own W12 addresses to node A by sending the addr_ack data packet, and updates its own address space and node type. After receiving the reply, node A updates its address table while obtaining the address. If there are redundant addresses, it will mark itself as a proxy node; if there are no redundant addresses, it will mark itself as a normal node.

If node A is within the range of two hops and does not receive a reply after sending the address request within the specified time, it will continue to send the request to expand the address allocation range. As mentioned above, this is the value of the expanded hop count R. The value of R is equal to the set maximum value of hops Rmax. When configuring the IP address of the 6LoWPAN network, it can be distinguished into four different types according to the various roles played by the nodes. See Sect. 3.1 for specific definitions. At the same time, all types of nodes have their own address tables, focusing on recording their addresses and address spaces. When recording, new nodes or normal nodes only record their IP addresses, but before the new node has a unique address, its temporary address is replaced by the node's MAC address. The address table of the proxy node records its own spare address set and the number of addresses, as well as its own address. If there are consecutive addresses, it is required to synchronously record the number of consecutive addresses at the lower limit (start address) of the consecutive address set.

4.2 Address Recycling

The nodes of the 6LoWPAN network will use batteries for energy supply. Because the battery has a certain storage capacity and is affected by environmental factors such as floating wireless signals, the network nodes are often disconnected for different reasons. Therefore, before the node is disconnected, its address must be recovered. In a dynamic network, the head nodes, normal nodes, and proxy nodes will send addr_leave packets to notify neighbor nodes within one hop before leaving the network. The neighbor node that receives the addr_leave data packet will immediately send back the addr_hello data packet to the leaving node. The disconnected node will theoretically receive the reply information from all neighbor nodes, and send the addr_ok data packet to the first replying node. The data packet contains its own address information, and the nearest node will be notified through the data packet. At the same time, after the disconnected node receives the first reply, it stops receiving information and cannot continue to receive addr_hello packets from other nodes.

5 Experiment Analysis

The experiment uses OPNET Modeler as the simulation platform to simulate and analyze the proposed binary tree address allocation, and compare it with the DAAM address configuration scheme, mainly based on the aspects of the average overhead of address configuration and the success rate of address configuration. The nodes are randomly and evenly distributed in the simulation area of 200 m * 200 m square area. The network is divided into 8 network grids, that is, the ordinate and abscissa of the node occupy 2 bits respectively. The value of the number of nodes N is 40, 50, 60, 70, and 80 (Figs. 5, 6 and 7).

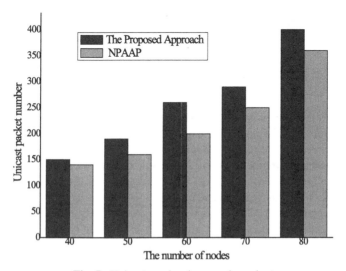

Fig. 5. Unicast overhead comparison chart

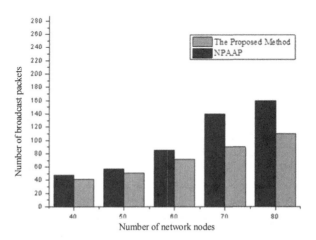

Fig. 6. Broadcast cost comparison chart

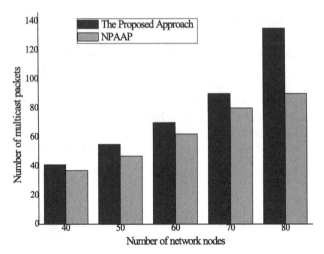

Fig. 7. Multicast overhead comparison chart

In the 6LoWPAN network, nodes within the communication range of this node can communicate directly, and nodes not within the communication range need to be forwarded by an intermediate node. The unicast propagation method is only carried out between two nodes, the multicast propagation method is that multiple nodes participate, and the scope of the broadcast is all nodes in the entire network. In the entire address configuration process, there are unicast and multicast, and broadcast transmission methods. In order to comprehensively compare and analyze the performance of the proposed solutions, the broadcast overhead, unicast overhead and multicast overhead of the NPAAP and the address configuration process proposed in this paper are compared and analyzed. The main statistical method is to set global variables during address configuration. Each time a type of data packet is sent, the number of corresponding information packets is increased.

Compared with NPAAP, the binary tree address configuration scheme proposed in this paper is mainly unicast overhead and multicast overhead. First, the network is divided into network grids based on location information to achieve clustering. Therefore, address configuration is performed in each grid, and there is almost no network-wide broadcast overhead. The new node only sends multicast packets to neighboring nodes within one hop when it joins the network and when the node leaves the network. If there is no free address within one hop, the node will expand the range of sending address requests. Therefore, only a limited number of multicast data packets are generated. At the same time, the range of multicast packets is generally limited to each grid. Although the number of multicast packets is large, the overhead of each multicast data packet is smaller than NPAAP. In addition, new nodes choose proxy nodes for themselves, which are generally close to them, so unicast overhead is also reduced compared with NPAAP.

6 Conclusion

In the binary tree address configuration mechanism studied in this paper, the proxy nodes in the grid usually obtain the corresponding address space automatically according to the number of IP address nodes in the vicinity that have not yet been allocated. This greatly reduces the failure probability of address configuration due to uneven node distribution, and thus reduces the frequency of starting the address borrowing mechanism, which decreases the consumption of address configuration to a certain extent. At the same time, the task of address configuration is assigned to the vast majority of nodes in the network according to a certain mechanism, instead of being concentrated in a small number of nodes, so as to avoid the problem of address configuration failure due to the death of the proxy node. The performance comparison with the traditional address configuration mechanism proves the feasibility of the location information-based address configuration mechanism.

References

1. Bocchino, S., Petracca, M., Pagano, P., et al.: SPEED-3D: a geographic routing protocol for 6LoWPAN networks. Int. J. Ad Hoc Ubiquit. Comput. **19**(3/4), 143–156 (2015)
2. Kim, E., Kaspar, D, Gomez, C., et al.: Problem statement and requirements for IPv6 overLow-power wireless personal area network (6LoWPAN) routing. Heise ZeitschriftenVerlag **15**(4), 26, 28, 30 (2012)
3. Hu, S.C., Lin, C.K., Tseng, Y.C., et al.: Distributed address assignment with address borrowing for Zig Bee networks. In: ICC - 2014 IEEE International Conference on Communication Workshop, pp. 454–459. IEEE (2014)
4. Droms, R., Bound, J., Volz, B., et al.: Dynamic host configuration protocol for IPv6 (DHCPv6). IEEE Syst. J. **21**(8), 67–81 (2003)
5. Jinmei, T.: RFC 2462: IPv6 stateless address autoconfiguration. IEEE Syst. J. **18**(8), 117–124 (1998)
6. Kassem, M.M., Hamza, H.S., Saroit, I.A.: A clock skew addressing scheme for internet of things. In: IEEE, International Symposium on Personal, Indoor, and Mobile Radio Communication, pp. 1553–1557. IEEE (2015)
7. Wang, X., Le, D., Yao, Y., et al. Location-based mobility support for 6LoWPAN wireless sensor networks. J. Netw. Comput. Appl. **49**(C), 68–77 (2015)
8. Montavont, J., Cobarzan, C., Noel, T.: Theoretical analysis of 1Pv6 stateless address autoconfiguration in low-power and Lossy Wireless Network. In: IEEE Rivf International Conference on Computing & Communication Technologies - Research, Innovation and Vision for the Future, pp. 198–203. IEEE (2015)
9. Jia, H., Gai, Y., Xu, D., et al.: Link importance-based network recovery for large-scale failures in smart grids. Wirel. Netw. **2020**, 1–13 (2020). https://doi.org/10.1007/s11276-019-02219-9

A Resource Consumption Attack Identification Method Based on Data Fusion

Libin Jiao[1], Yonghua Huo[1], Ningling Ge[2], Zhongdi Ge[3], and Yang Yang[3(✉)]

[1] Science and Technology on Communication Networks Laboratory, The 54th Research Institute of CETC, Shijiazhuang, Hebei, People's Republic of China
[2] R & D Department, Agricultural Bank of China, Shanghai, People's Republic of China
[3] State Key Laboratory of Networking and Switching Technology, Beijing University of Posts and Telecommunications, Beijing, People's Republic of China
yyang@bupt.edu.cn

Abstract. Data fusion can make use of information from different sources or different representations to describe the target more accurately, which has important research significance. Aiming at the network-running node may be attacked or there is measurement error, this paper comprehensively utilizes the information of each node, and proposes a resource consumption attack identification method based on node multi-dimensional data fusion. First, construct a correlation matrix between nodes, identify normal nodes and possible abnormal nodes, and assign different weights to each node. Then, calculating the support of the node's system attributes for the attack type, and adopting the D-S evidence theory to effectively identify the network attack. The simulations demonstrate the effectiveness and certain advantages of the proposed algorithm.

Keywords: Resource consumption attack · Correlation analysis · Data fusion · D-S evidence theory

1 Introduction

The ultimate threat of various attacks or normal behavior peaks is the availability of the network-operating environment, and the availability of the network depends mainly on the occupancy of various network resources, such as network bandwidth, throughput, storage and computing resources. Network resource consumption attacks generally refer to hackers using reasonable service requests to preempt excessive system resources, so that other normal users cannot obtain sufficient resources, causing the system to stop responding to service requests. In the case of a terminal server, since resources such as bandwidth, computing power, and storage have certain limitations, when a hacker generates an excessive number of network server requests, a large amount of resources of the terminal server are abused by the terminal, and normal users cannot use the service. Resource-consumption attacks are always accompanied by anomalies in the system attributes of each node, while nodes do not exist independently in the network environment, and the attribute changes of each node have potential relevance.

© ICST Institute for Computer Sciences, Social Informatics and Telecommunications Engineering 2021
Published by Springer Nature Switzerland AG 2021. All Rights Reserved
M. Cheng et al. (Eds.): SmartGIFT 2020, LNICST 373, pp. 128–139, 2021.
https://doi.org/10.1007/978-3-030-73562-3_11

Data fusion is a multi-level and multi-faceted processing process, which involves detecting, correlating, combining and estimating data to improve the estimation accuracy of states and characteristics, and then obtain a more accurate description of the perceived objects. The current data fusion methods can be roughly divided into two categories: probabilistic statistics methods and artificial intelligence methods. Among them, the probabilistic method mainly has the following types: the first is the weighted average method using the simplest and most intuitive mathematical operation fusion; the second is the Kalman filter method used for real-time fusion of dynamic low-level redundant data; the third is the multi-Bayesian estimation which minimizes the likelihood function of the associated probability distribution function; the fourth is the D-S evidence theory applicable to the inference of uncertain problems. The artificial intelligence mainly includes: fuzzy logic theory based on multi-valued logic reasoning, combined calculation based on fuzzy set theory; neural network method using data processing capability and automatic reasoning ability of neural network to realize data fusion and so on. Each of these methods has its application, and D-S evidence theory can express the correct, incorrect and uncertain probabilities at the same time, and has certain advantages in dealing with the uncertainty of data.

Therefore, this paper proposes a resource consumption attack identification method based on node multi-dimensional data fusion. The research contents and innovations of this paper are as follows:

(1) According to the correlation analysis, construct a correlation matrix between nodes to distinguish between normal nodes and nodes that may be attacked. Then, using the system attribute information of the node, the node data is transformed into the evidence in the D-S evidence theory.
(2) Using the D-S evidence theory, the individual evidence is distributed according to the basic probability of the node, and the network resource consumption attack is effectively identified.

2   Related Work

Network attacks can be divided into two categories: 1) Using information system security vulnerabilities to bypass information system security protection measures and enter information systems to achieve the purpose of controlling information systems. Such attacks are generally called control attacks. 2) Although such a network attack cannot control the information system, the information system's service capability is degraded or the service capability is completely lost due to the large consumption of information system resources, such as memory resources, computing resources, bandwidth resources, and the like. Such attacks are generally called resource consumption attacks. The resource consumption attack is mainly to exhaust a certain network resource. In the attack process, the terminal server attacks the terminal server, and the terminal server's CPU, memory, communication link and other resources are overloaded. The traditional means of network resource consumption attack countermeasures are intrusion detection and intrusion response. For example, the detection method based on anomaly detection can be used for detecting network resource consumption type attacks.

Palmieri F et al. [1] proposed a two-stage anomaly detection strategy, using independent component analysis modeling as a blind source separation problem, and constructing a baseline traffic distribution, thereby transforming network attack detection into anomalous/normal classification problem. Aborujilah A et al. [2] focused on the impact of DoS attacks on CPU power performance and network bandwidth. In order to evaluate CPU and bandwidth power performance, real flood attacks were implemented in different scenarios. Harshaw C R et al. [3] proposed a graph parsing method for detecting network stream data anomalies, which represents a time slice of traffic as a graph, and identifies an abnormal interval by performing anomaly detection on the graph primitive count sequence.

Palmieri F et al. [4] proposed a network-based anomaly detection method based on the analysis of non-fixed attributes and the "hidden" recurrence pattern that occurs in aggregated IP traffic flows. Recursive quantitative analysis was used to explore the hidden dynamics and temporal correlation of statistical time series. A Chaudhary et al. [7] used the ability to deeply learn the topological features of social networks to detect anomalies in email networks and Twitter networks, and proposed a model neural network model and applied it to social contact graphs. Considering the combination of various social network statistical measures, the structure and function of the abnormal nodes are studied by using deep neural networks on them, which found that the hidden layer of the neural network plays an important role in discovering the impact of statistical metric combinations in anomaly detection.

X. Chun-Hui et al. [8] proposed an adaptive method based on the iForest algorithm. Some feature extractors are constructed by statistical methods to highlight different anomalous behaviors in different indicators, and then the extracted feature data is used for iForest construction and prediction. Combined with a specific feature extractor, you can eliminate periodic effects or specify peaks or valleys to accommodate different metrics. Rapid detection of large data sets with the iForest algorithm with linear time complexity and low memory requirements. R. Liu et al. [9] proposed a network anomaly detection method (NAD-NNG) based on the idea of natural neighborhood graph. In order to eliminate noise points or mark erroneous points and reduce the time complexity of anomaly detection, the algorithm clusters the normal data set using a natural neighborhood map, and adaptively obtains a percentage value β for setting an abnormal threshold.

Q. Su et al. [10] proposed a genetic algorithm based on Management Information Base (MIB) to detect network anomalies. The algorithm is based on the classification theory using integrated IF-THEN rules, proposes a new chromosome-coding scheme, and discusses a new charging function design method. Ç. Ateş et al. [11] proposed a new method for network anomaly detection based on the probability distribution of header information. The Greedy algorithm is used to calculate the distance between the header distributions to reflect the main features of the network, eliminating some of the requirements associated with Kullback-Leibler divergence. The support vector machine classifier is then used during the detection phase to reduce the false alarm rate and adapt the system to different networks.

The traditional attack detection technology focuses on system intrusion detection, anti-virus software or firewall of the user network. The biggest problem is that when the attack data stream reaches a peak, a large amount of data is collected on the victim side, and there is no effective way to filter the data. Thus, the user network can only be in a state of passive defense. In order to effectively prevent attacks on the victim side, it is necessary to find the attack behavior at an early stage. However, it is difficult to distinguish normal users and attack data in the early stage of the attack, which brings difficulties for resource consumption attack identification. In addition, once the attack occurs, whether the malicious data seriously affects the victim, the data has been transmitted in the network, causing actual waste of network resources. Because resource consumption attack types are complex and versatile, it is unrealistic to detect whether a single packet is a malicious attack. Therefore, we analyze the trend of node resources in the network and the correlation between them to achieve early detection and identification of resource consumption attacks.

3 Data Fusion Algorithm

3.1 Algorithm Flow

This paper proposes a resource consumption attack identification method based on node multi-dimensional data fusion. According to the resource consumption type attacks that may exist in the network, the correlation matrix between the nodes is constructed according to the correlation analysis, and the normal nodes and the nodes that may be attacked are distinguished. Then, the system attribute information of the integrated node is used to convert the node data into evidence in the D-S evidence theory. Finally, using the D-S evidence theory, each evidence is distributed according to the basic probability of the node, so as to effectively identify the resource-consuming attack. The specific steps of the algorithm are as follows, as shown in Fig. 1.

Input: system attribute data of the node.
Output: resource consumption attack identification result.
Step 1: construct a correlation matrix of nodes by using the maximum information coefficient, and identify possible abnormal nodes;
Step 2: Integrate all node information, convert the node data into evidence in D-S evidence theory, and assign different weights to different evidences to describe the credibility of each evidence;
Step 3: According to the basic probability allocation BPA evaluates the source of evidence, so as to effectively identify the resource-consuming attack.

3.2 Basic Probability Allocation Based on Correlation Analysis

According to the possible existence of attack behavior in the network, the correlation matrix between nodes is constructed according to the correlation analysis to distinguish the normal node from the node that may be attacked. Then, the system attribute information of the node is comprehensively utilized to convert the node data into evidence. Different weights are assigned to describe the credibility of each evidence. The treatment of conflict evidence is to assign smaller weights instead of removing conflict evidence,

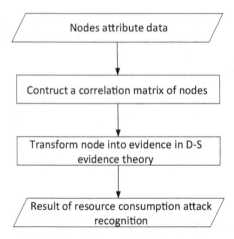

Fig. 1. Overall flowchart of the algorithm

mainly for two reasons: (1) When the amount of evidence is small, it is impossible to determine which evidence is conflict evidence, and appropriate weight can reduce the negative impact of conflict evidence on the fusion result. When the number of evidence is large, the weight of the conflict evidence will be approximately zero, and its impact on the fusion result is negligible. (2) Unless it is impossible to determine which evidence is conflict evidence or which raw data is erroneous data before knowing the correct decision, and because of the universality of noise in the data, the fusion rule must be able to adapt to the existence of conflict evidence.

Suppose the network has a certain number of nodes, each of which has a series of system attributes such as CPU utilization, latency, bandwidth, packet loss rate, and number of connections. First, the correlation between nodes is calculated using the maximum information coefficient, thereby constructing a node correlation matrix. Note that the attribute of a node s_i at a certain time is a random variable s_X, and the attribute of the other node s_j in the same time is a random variable s_Y, so that a finite data set D is formed between any two nodes, $D = \{s_X, s_Y\}$. The correlation of nodes s_i and s_j can be calculated as follows:

$$Sim(s_i, s_j) = \max_{a*b<B}\{M(D)_{a,b}\} = \max_{a*b<B}\left\{\frac{I^*(s_X, s_Y)}{\log \min\{a, b\}}\right\} \tag{1}$$

Where $1 < B \leq N^{1-\varepsilon}$, $0 < \varepsilon < 1$. In general, $\varepsilon = 0.4$ can be used to obtain better results, so $B = N^{0.6}$ is taken in this paper. $a * b$ represents a different division of the data set D, and $I(s_X, s_Y)$ represents mutual information of the random variables s_X and s_Y.

$$I(s_X, s_Y) \approx I_{D|G}(s_X, s_Y) = \sum_{j=1}^{b}\sum_{i=1}^{a} P_{(x,y)}(i, j) \log \frac{P_{(x,y)}(i, j)}{P_x(i)P_y(j)} \tag{2}$$

Where $P_x(i) \approx \frac{n_x(i)}{N}$, $P_y(j) \approx \frac{n_y(j)}{N}$, $P_{(x,y)}(i, j) \approx \frac{n_{(x,y)}(i,j)}{N}$, $n_x(i)$ is the number of samples falling into the i-th division grid of the random variable s_X, $n_y(j)$ is the number of samples falling into the j-th division grid of the random variable s_Y, $n_{(x,y)}(i,j)$ is the number of samples that fall into the i-th division grid of the random variable s_X and the j-th division grid of the random variable s_Y at the same time.

According to the correlation matrix, the normal node set and the possible abnormal node set are distinguished, which are respectively recorded as S_0 and S_1.

$$Sim(s_i) = \frac{1}{n-1} \sum_{j=1}^{n-1} Sim(s_i, s_j) \tag{3}$$

$$\overline{Sim} = \frac{1}{n} \sum_{i=1}^{n} Sim(s_i) \tag{4}$$

Where $Sim(s_i)$ represents the overall correlation of the node s_i in the network, n represents the number of nodes in the network, and $\overline{Sim}$ represents the average correlation of all nodes in the network. Put all the nodes of $Sim(s_i) \geq \overline{Sim}$ into the set S_0 and mark them as normal nodes, put all the nodes of $Sim(s_i) < \overline{Sim}$ into the set S_1 and mark them as possible abnormal nodes.

Then, the set of node system attributes associated with each resource-consuming attack type is recorded as $V = \{V_1, V_2, \cdots, V_k\}$. The evidence generated by each node is calculated for the mass function generated by different attack types as follows:

$$m(s_i) = \sum_{j=1}^{k} \beta_{ij} \tag{5}$$

$$\beta_{ij} = \frac{V_i^j - \overline{V_j}}{V_{max}^j - V_{min}^j} \tag{6}$$

Where β_{ij} represents the support degree of the j-th system attribute of each node to the attack type, V_{min}^j and V_{max}^j respectively represent the minimum and maximum values of the j-th system attribute of the normal node set S_0, $\overline{V_j}$ denotes the average value of the j-th system attribute of the set S_0, and V_i^j represents the j-th system attribute value in the node s_i that is higher than the average of the set S_0. Because the resource consumption attack behavior is mainly to maliciously seize the system resources, the node attribute value is too high, so this paper only considers the attribute higher than the average value of the normal node.

So far, n pieces of evidence are generated by n nodes. The resource usage anomaly caused by the resource consumption attack behavior is mainly manifested in the system attribute of the abnormal node, and the normal node has a relatively small effect on the attack type identification. In order to further improve the fusion precision, the weight of the evidence generated by each node will be calculated and used as the correction coefficient of the evidence.

$$w_i = \frac{\sum_{i=1}^{n} Sim(s_i)}{Sim(s_i)} \tag{7}$$

Finally, using the correction factor to normalize the weighted correction of each evidence, the BPA of the revised evidence is as follows:

$$m'(s_i) = \frac{w_i m(S_i)}{\sum_{p=1}^{N} w_p m(S_p)} \tag{8}$$

The evidence theory was first proposed by Dempster, who gave the concept of upper and lower probabilities and gave the principle of synthesis of two independent sources of information. Later, his student Shafer further improved and perfected his theory. Therefore, the evidence theory is also called Dempster-Shafer evidence theory, referred to as D-S evidence theory. The core content of D-S evidence theory is "evidence" and "combination". "Evidence" refers to data containing uncertain information. "Combination" refers to the synthesis rule. The synthesis formula can combine the information represented by the data to get more reliable and effective conclusions, which makes D-S evidence theory widely used in many fields such as financial analysis and intelligence analysis.

The following introduces the basic concepts of D-S evidence theory:

(1) Identification framework

Suppose there is a finite set of non-empty hypotheses Θ as the identification framework for evidence theory, consisting of N mutually exclusive hypotheses, defined as:

$$\Theta = \{H_1, H_2, \cdots, H_N\} \tag{9}$$

Where N is the number of hypotheses in the recognition system, H is to identify each hypothesis in the system, and all decision plan sets made by the system are a subset of the power set 2^{Θ} of the identification framework Θ.

(2) Basic probability allocation

The basic probability allocation (BPA) under the identification framework Θ is a function under the mapping $m: 2^{\Theta} \rightarrow [0, 1]$, which satisfies the following constraints:

$$\begin{cases} m(\emptyset) = 0 \\ \sum_{A \subseteq \Theta} m(A) = 1 \end{cases} \tag{10}$$

Where A is a proposition containing one or more hypotheses in the identification framework Θ, $m(A)$ represents the degree of support of the evidence for Proposition A, and any proposition A that satisfies $m(A) > 0$ is called a focal element.

(3) Belief function

The belief function (Bel) of Proposition A indicates the degree of trust in the event that Proposition A is true. The trust function under the identification framework Θ is defined as:

$$\text{Bel}(A) = \sum_{B \subseteq A} m(B) \quad A, B \subseteq \Theta \tag{11}$$

Among them, A and B are propositions in 2^{Θ}, and m is the basic probability distribution function on Θ. If an interval is used to indicate the strength of support for any one proposition, then the belief function is the lower bound of this interval.

(4) Plausibility function

The plausibility function (Pl) under the recognition framework Θ is defined as:

$$Pl(A) = \sum_{B \cap A \neq \emptyset} m(B) \qquad A, B \subseteq \Theta \tag{12}$$

Among them, A and B are propositions in 2^{Θ}, and m is the basic probability distribution function on Θ. If an interval is used to indicate the strength of support for any one proposition, then the belief function is the upper bound of this interval.

(5) Synthetic rule

Assuming that m_1 and m_2 are two independent basic probability distribution functions defined on the identification framework, set $A, B, C \subseteq \Theta$, then the Dempster synthesis rule is defined as:

$$m(A) = \begin{cases} \frac{\sum_{B \cap C = A} m_1(B) \cdot m_2(C)}{1-k}, & A \neq \emptyset \\ 0, & A = \emptyset \end{cases} \tag{13}$$

Where k is the evidence conflict factor. When $k = 1$, the evidence completely conflicts, and the denominator of the synthetic rule formula is 0, and the synthesis rule loses its meaning. In the case of $0 < k < 1$, the basic probability allocation of Proposition B and Proposition C can be fused using a synthesis rule.

Because the evidence theory can deal with the uncertainty of the data well, and can express the correct, incorrect and uncertain probabilities at the same time, it is very suitable for the identification of network resource-consuming attacks. Based on evidence-based fusion formulas, one or more sets of evidence can be combined into a new piece of evidence.

3.3 Resource Consumption Attack Identification Method Based on Data Fusion with D-S Theory

Since the nodes in the network may not only be subjected to various attacks, the data may have different degrees of measurement errors. Therefore, the attack behavior needs to be identified for subsequent better prevention. In this paper, the D-S evidence theory is used to distribute and fuse each evidence according to the basic probability of nodes to effectively identify network resource consumption attacks.

First, the BPA evaluates the source of evidence according to the basic probability, and calculates the trust function and likelihood function of each proposition; Secondly, the interval number is constructed by the trust function and the likelihood function to obtain the trust interval of each proposition; Finally, using the D-S evidence theory, the individual evidence is distributed according to the basic probability of the node, by sorting the results of the aggregation, the recognition result of the resource-consuming attack is obtained.

4   Simulation and Analysis

The simulation uses the public dataset to verify the algorithm, which is compared with the classical D-S evidence theory algorithm [12] and the recent improved algorithm [13, 14]. It proves that our algorithm is effective in solving the evidence conflict problem and has a high correct rate. The experimental algorithm is written in Python language and implemented in the operating environment of ASUS notebook (CPU 1.80GHz, memory 8GB, hard disk 512GB SSD, Win10 operating system).

(1)   Public datasets

In order to verify the accuracy and validity of the algorithm fusion results, the algorithm is validated by Iris dataset [5] and KDD CUP 99 dataset [6]. The Iris dataset contains 150 data, divided into 3 categories, 50 data per category, and each data contains 4 attributes. Use the attributes of flower length (SL), flower width (SW), petal length (PL), and petal width (PW) to predict the flower belongs to which category. The KDD CUP 99 dataset is a nine-week network connection data collected from a simulated US Air Force LAN, containing one normal identification type and 22 training attack types. Each connection record contains 41 fixed feature attributes and a class identifier. The identifier is used to indicate whether the connection record is normal or a specific attack type (Table 1).

Assuming the identification framework of the Iris data set $\Theta = \{S, E, V\}$, the existing four evidences are the attribute length (SL), the width of the flower (SW), the length of the petal (PL), and the width of the petal (PW), which BPA are expressed as follows:

Table 1. The BPA of different resources

Resource	BPA
SL	$m_1(S) = 0.41, m_1(E) = 0.29, m_1(S, E) = 0.3$
SW	$m_2(S) = 0.58, m_2(E) = 0.07, m_2(S, V) = 0.35$
PL	$m_3(S) = 0.3, m_3(E) = 0.15, m_3(S, E) = 0.2, m_3(S, V) = 0.35$
PW	$m_4(S) = 0.2, m_4(E) = 0.3, m_4(S, V) = 0.5$

It can be seen that all the evidence supports the proposition S is relatively large, so the fusion result should support the proposition S. The comparison between the classic D-S improved algorithm and the recent improved algorithm and the proposed algorithm fusion result is shown in Fig. 2. All algorithms have the highest support for Proposition S. Among them, the support of the proposed algorithm for the propositions S, E and V are 0.6261, 0.2023 and 0.1716 respectively, and the support of the proposition S is significantly higher than the proposition E and the proposition V. Moreover, the support of our algorithm for correct proposition S is 46.08%, 30.11% and 12.39% higher than that of Yager, Zhao and Jiang respectively. It can be seen that the algorithm of this paper can correctly process the Iris data set.

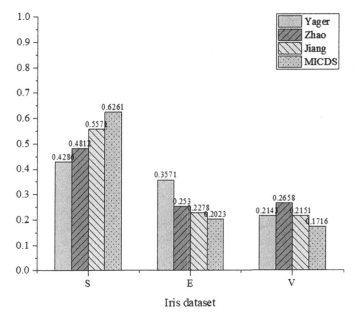

Fig. 2. Iris dataset fusion result

In addition, the accuracy of data fusion results relative to real results (Accuracy) is used as the evaluation index of the algorithm. Each experimental result is obtained from the average of 10 data sets each consisting of 100 sample points. The accuracy of fusion of abnormal points at different scales is shown in Fig. 3 and Fig. 4. As shown in Fig. 3, when there is no abnormal evidence, the accuracy of the algorithm is 89.3%, 92.5%, 93.1%, 95.7%. The accuracy of our algorithm is the highest, which is 7.16%, 3.46% and 2.80%

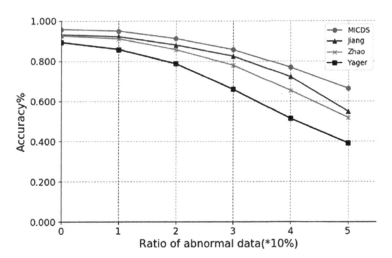

Fig. 3. Accuracy of the Iris dataset

higher than the other three algorithms. The accuracy of the four algorithms gradually decreases with the increase of the proportion of abnormal evidence, but the decrease of our algorithm is the slowest. As shown in Fig. 4, when all data are normal, the algorithm accuracy rates respectively are 83.1%, 87.5%, 89.1%, and 92.7%. The accuracy of the four algorithms decreases with the increase of the proportion of abnormal evidence, but our algorithm has the slowest decline rate. When mixed with 50% abnormal evidence, the accuracy rate remains at 61.6%, which is 64.7%, 24.9% and 14.9% higher than other algorithms. It can be seen that our algorithm can better deal with abnormal evidence and obtain higher recognition accuracy.

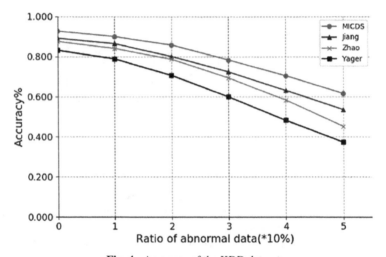

Fig. 4. Accuracy of the KDD dataset

5 Conclusion

Aiming at the network running node may be attacked or there is measurement error, this paper comprehensively utilizes the information of each node, and proposes a resource consumption attack identification method based on node multi-dimensional data fusion. Using correlation, the normal nodes and possible abnormal nodes are divided, and each node is assigned different weights, thus converting the nodes into evidence in D-S evidence theory. Then, according to the basic probability distribution function of the node to the attack type, the D-S evidence theory is used for fusion, and the resource consumption type attack is effectively identified. The simulation is carried out on the public datasets. The effectiveness and certain advantages of the proposed algorithm are proved by comparing the correctness of the algorithm and the comparison algorithm. The algorithm in this paper can realize the identification of resource-consuming attack types, but its fusion accuracy rate needs to be improved. The next step is to conduct in-depth research on this aspect.

Acknowledgement. This work was supported by National Key R&D Program of China (2019YFB2103202, 2019YFB2103200), Open Subject Funds of Science and Technology on Communication Networks Laboratory (6142104200106).

References

1. Palmieri, F., Fiore, U., Castiglione, A.: A distributed approach to network anomaly detection based on independent component analysis. Concurr. Comput. Pract. Exp. **26**(5), 1113–1129 (2014)
2. Aborujilah, A., Musa, S.: Detecting TCP SYN based flooding attacks by analyzing CPU and network resources performance. In: International Conference on Advanced Computer Science Applications & Technologies. IEEE (2015)
3. Harshaw, C.R., Bridges, R.A., Iannacone, M.D. et al.: GraphPrints: towards a graph analytic method for network anomaly detection (2016)
4. Palmieri, F., Fiore, U.: Network anomaly detection through nonlinear analysis. Comput. Secur. **29**(7), 737–755 (2010)
5. https://archive.ics.uci.edu/ml/datasets/Iris/
6. https://kdd.ics.uci.edu/databases/kddcup99/kddcup99.html
7. Chaudhary, A., Mittal, H., Arora, A.: Anomaly detection using graph neural networks. In: 2019 International Conference on Machine Learning, Big Data, Cloud and Parallel Computing (COMITCon), Faridabad, India, pp. 346–350 (2019)
8. Chun-Hui, X., Chen, S., Cong-Xiao, B., Xing, L.: Anomaly detection in network management system based on isolation forest. In: 2018 4th Annual International Conference on Network and Information Systems for Computers (ICNISC), Wuhan, China, pp. 56–60 (2018)
9. Liu, R., Zhu, Q.: A network anomaly detection algorithm based on natural neighborhood graph. In: 2018 International Joint Conference on Neural Networks (IJCNN), Rio de Janeiro, pp. 1–7 (2018)
10. Su, Q., Liu, J.: A network anomaly detection method based on genetic algorithm. In: 2017 4th International Conference on Systems and Informatics (ICSAI), Hangzhou, pp. 1029–1034 (2017)
11. Ateş, Ç., Özdel, S., Yıldırım, M., Anarım, E.: Network anomaly detection using header information with greedy algorithm. In: 2019 27th Signal Processing and Communications Applications Conference (SIU), Sivas, Turkey, pp. 1–4 (2019)
12. Yager, R.R.: On the aggregation of prioritized belief structures. IEEE Trans. Syst. Man Cybern. Part A: Syst. Hum. **26**(6), 708–717 (2002)
13. Zhao, Y., Jia, R., Shi, P.: A novel combination method for conflicting evidence based on inconsistent measurements. Inf. Sci. **367**, 125–142 (2016)
14. Jiang, W.: A correlation coefficient for belief functions. Int. J. Approx. Reason. **103**, 94–106 (2018)

Author Index

Cai, Hao 65
Chen, Fang 14
Chen, Qi 3
Chuan, Liu 107

Dai, Yong 55
Dan, Luo 117
Deng, Yuanshi 3
Duan, Kai 42

Gao, Lifang 93
Ge, Ningling 128
Ge, Zhongdi 128
Guo, Jing 93

Han, Cao 117
Hao, Jiang 117
Huang, Jin 77
Huo, Yonghua 128

Jiang, Chunxia 65
Jiang, Song 65
Jianwei, Li 117
Jiao, Libin 128
Jing, Tao 107

Li, Bozhong 14
Li, Wei 55, 65
Li, Zhi 42
Li, Zhong 3
Li, Zifan 14
Liao, Boxian 93
Lin, Peng 55, 65

Liu, Tingfeng 93
Liu, Zhengyuan 14, 77
Lv, Junfeng 14

Mao, Junli 27
Miao, Weiwei 77

Ning, Xin 3

Shi, Linshan 77

Wang, Haoyu 27
Wang, Qicai 27
Wang, Xinglong 77
Wei, Donghong 27
Wu, Xilao 55

Xia, Ping 65
Xiaoyu, Wang 107
Xu, Lu 107, 117
Xu, Tingting 42
Xu, Yong 55

Yang, Huifeng 93
Yang, Junzhong 77
Yang, Yang 128
Yu, Peng 14, 27, 77

Zhang, Rui 3
Zhang, Xiaotao 93
Zhonghua, Liang 107
Zhu, Ke 3

Printed in the United States
by Baker & Taylor Publisher Services